Storm Water Pollution Prevention Plan Risk Management

FOR THE CONSTRUCTION INDUSTRY

Storm Water Pollution Prevention Plan Risk Management

FOR THE CONSTRUCTION INDUSTRY

Applying Sustainable Environmental Responsibility and Stewardship

Mike L. Peters QSD, CPESC

ISBN: 1508666423
ISBN 13: 9781508666424
Library of Congress Control Number: 2015903471
CreateSpace Independent Publishing Platform
North Charleston, South Carolina

Contents

Storm Water Pollution Prevention Plan (SWPPP) Risk Management

FOR THE CONSTRUCTION INDUSTRY

Understanding the reason, course, logic, and responsibility behind a Storm Water Pollution Prevention Plan.

Introduction

The acronym "SWPPP" has been around since at least the 1990s. It stands for Storm Water Pollution Prevention Plan. It has been referred to as a "Sweet Pea," "SWIPP," "SWEEP," or "SWEEPY," just to name a few. A SWPPP is actually a living, dynamic set of engineered drawings, certifications, substantive language, and descriptions set forth in a written plan—electronic or bound in binders—describing best management practices (BMPs), means, and methods for state, federal, and local storm water pollution law compliance to be implemented before, during, and after the completion of a construction project.

In the "good old days" of construction projects and manifest destiny, from the industrial revolution to the late 1970s, the Clean Water Act was unheard of for the most part. Burgeoning projects created disturbed soil areas (DSAs), allowing soil, silt, sediment, and pollutants to enter the water bodies of the United States (and world) in the name of industry and social, brick-and-mortar progress, resulting in the destruction, eradication, and irreparable or perpetual degradation of countless virgin ecosystems.

While performing hydro geological research and testing for a client in the mountains of West Virginia in 2007, I asked the local contact there how the fishing was in the nearby streams and rivers. His reply was that many of the streams were still "dead" or bereft of aquatic ecosystem life, for example, vertebrates and macro invertebrates, due to the effects of coal mining, pulp mill, and other industrial activity in the not so distant past.

Moving into the twenty-first century, it is essential as environmental stewards that we as a society and culture, both local and global, embrace care, concern, aware-ness, stewardship, and sustainability regarding storm water pollution prevention of our natural and renewable resources.

SWPPPs are developed and implemented to minimize the effects and impacts of soil disturbance and construction activity on the environment. Erosion is a natural process, but accelerated erosion or the introduction of volatile organic or hydrocarbon-based compounds into the water and soil due to human activity should be mitigated in order to minimize or inflict zero harm regarding preventable negative human impact on the natural environment.

There used to be a defined annual "rainy season" in the SWPPP world, but now, for the most part, SWPPP implementation must be performed/addressed year-round, especially pre, during, and post rain events. The defined "rainy season" used to be, depending on which state and what area of the state the project is in and which permits are applicable to the project, between the months of April 15 or May 15 to September 15 or October 15. Now SWPPP BMP implementation for water quality, soil stabilization, and sediment control is required year-round, with additional requirements pre, during, and post storm events for pH and turbidity discharge monitoring. Other best management practices (BMPs) for non-storm water or waste management-related requirements, such as material storage, spill control, fueling and equipment maintenance, work over water, solid and concrete waste management, and so on, are enforceable as required on a case-by-case basis, per the applicable permits for the project.

The main body of a SWPPP has six best management practices (BMP) sections:

1. Soil Stabilization BMPs (13)
2. Sediment Control BMPs (11)
3. Tracking Control BMPs (3)
4. Wind Erosion Control (1)
5. Non-Storm Water Site Management BMPs (15)
6. Waste Management (10)

These BMPs will be described briefly later in this publication.

Part One

Introduction to the Storm Water Pollution Prevention Plan (SWPPP)

CHAPTER 1

Why SWPPP? A Brief History

The Federal Water Pollution Control Act of 1948 gave birth to the Federal Water Pollution Control Act Amendments of 1972, which greatly expanded and defined the implementation of water pollution control requirements mandated by the federal government.

The key that changed storm water pollution laws forever was the 1972 Federal Water Pollution Control Act Amendments act, which introduced the National Pollutant Discharge Elimination System (NPDES) permit, which is a permit system for regulating point sources of pollution. A "point source" is a single identified source of origination of something.

Per a recent Wikipedia article on the Clean Water Act regarding point source pollution control precedence,

Point sources include:

- industrial facilities (including manufacturing, mining, oil and gas extraction, [7] and service industries).
- municipal governments and other government facilities (such as military bases), and
- some agricultural facilities, such as animal feedlots.

Point sources may not discharge pollutants to surface waters without a permit from the National Pollutant Discharge Elimination System (NPDES). This system is managed by the United States Environmental Protection Agency (EPA) in partnership with state environmental agencies. EPA has authorized 46 states to issue permits directly to the discharging facilities. The CWA also allows tribes to issue permits, but no tribes have been authorized by EPA. In the remaining states and territories, the permits are issued by an EPA regional office.[8] (See Titles III and IV.)

In previous legislation, Congress had authorized states to develop water quality standards, which would limit discharges from facilities based on the characteristics of individual water bodies. However, these standards were only to be developed for interstate waters, and the science to support this process (i.e. data, methodology) was in the early stages of development. This system was not effective and there was no permit system in place to enforce the requirements. In the 1972 CWA Congress added the permit system and a requirement for technology-based effluent limitations. (http://en.wikipedia. org/wiki/Clean_Water_Act)

The same article in Wikipedia reveals the precedent for technology-based standards:

The 1972 CWA created a new requirement for technology-based standards for point source discharges. EPA develops these standards for categories of dischargers, based on the performance of pollution control technologies without regard to the conditions of a particular receiving water body. The intent of Congress was to create a "level playing field" by establishing a basic national discharge standard for all facilities within a category, using a "Best Available Technology." The standard becomes the minimum regulatory requirement in a permit. If the national standard is not sufficiently protective at a particular location, then water quality standards may be employed. (http://en.wikipedia.org/wiki/Clean_Water_Act, January 10, 2015)

These 1972 precedent-setting changes in federal and state water pollution control practices were the beginning framework for establishing the Clean Water Act of 1977 and the Water Quality Act of 1987, the need for storm water pollution control compliance, and the documentation of that compliance from point source dischargers.

Section 401 of the Clean Water Act required applicants to file for federal permits in order to conduct any activity, including the construction or operation of a facility that may create or result in a discharge of any pollutant. Dischargers must obtain certification of those activities from the state where the point source discharge originates. This is known as a "Water Quality Certification."

Section 402 of the Federal Water Pollution Control Act established the National Pollutant Discharge Elimination System (NPDES) permitting program. This program required the control and documentation of discharges of pollutants from point source dischargers and the elimination of such discharges as emphasized by the implementation of technology-based best management practices (BMPs).

Receiving Waters

Section 402 was also the beginning of a series of studies collectively known as the Nationwide Urban Runoff Program (NURP). One of the main purposes of NURP was to differentiate natural storm water runoff from Municipal Separate Storm Sewer Systems (MS4s) from storm water runoff (discharged) from construction and industrial activities. MS4 drainage systems can be any number of drainage conveyance systems, including, but not limited to, culverts (concrete, composite, or metal), human-made drainage swales or ditches, or natural swales or drainage ditches that convey storm water discharged from construction or industrial sites, to "waters of the United States."

"Waters of the United States" are known as, but not limited to, waters that may be used in interstate or foreign commerce, including waters subject to tidal influence, wetlands, lakes, rivers, streams (including ephemeral, seasonal streams), mudflats, wet meadows, playa flats, ponds, the territorial sea, and waste treatment systems including ponds or lagoons designed to meet the requirements of the Clean Water Act.

Additionally, wetlands can be, but are not limited to, areas that are affected by surface water or groundwater with a frequency or duration that is sufficient to support vegetation that is typically adapted for life in saturated soil conditions.

At the present day, there is a differentiation between naturally occurring storm water runoff and storm water runoff that is affected by industrial, human-made, or construction activities. "Only rain down the drain" is the new mantra for environmental awareness. No sediment, particulates, chemicals, or human-made products that will negatively or deleteriously affect aquatic life, ecosystems, or drinking water aquifer quality shall be discharged to a receiving water body or MS4 drainage under the Clean Water Act.

CHAPTER 2

National Pollutant Discharge Elimination System (NPDES) Permitting

Phase I NPDES permitting construction activities are operations that result in the disturbance of five acres or more of total land area that is subject to regulation. Phase II construction sites are between one to less than five acres of disturbed area that is subject to regulation.

These Phase I and Phase II projects are required to apply for an NPDES permit prior to soil disturbing activities. The operator, "owner," or "Legally Responsible Person" (LRP, must be a person representing him or herself and/or a legal entity (corporation or group)) must file for permit coverage by filing a Notice of Intent (NOI) at least seven days before beginning construction.

As of 2010, forty states and territories had NPDES permitting authority, and thirteen states and territories did not have NPDES authority. In states without NPDES authority, the program is promulgated by the regional office of the EPA. Any projects in states without NPDES authority will use the application methods approved by their regional EPA office. EPA's general construction permit requires that projects and facilities operate under federal, state, or local SWPPP compliance or storm water management plans, where Notices of Intent must be signed and submitted to the state or local agency prior to soil disturbing activities.

These permits will be filed with the controlling entity of the state and/or federal government. Typically, the state or federal Environmental Protection Agency, Department of Environmental Quality, or Regional Water Quality Control Board will be involved in the promulgation of the permit. Additionally, the Department of Fish and Wildlife, State/Federal Parks System, US Army Corps of Engineers, and other local or national agencies may also be involved in the permitting process.

It is of the utmost importance that all of the applicable regulatory permits are addressed in the NOI and the SWPPP to avoid scheduling and permitting compliance issues. On many projects tree removal must be performed in the winter months of the year, prior to bird migration and nesting activities. If wildlife or environmental permit task windows are missed, the entire project may be delayed until the next calendar year or longer. It is a good idea to place all environmental and fish and wildlife windows of activity and nonwork windows on the project schedule so that the applicable permit restrictions are on the schedule radar and do not end up delaying the project.

The same rules apply for obtaining right-of-way (ROW) permits for project operation use and utility installation or removal as applicable. Right-of-way delays, such as power pole relocation or tree removal, can delay your project for months or seasons.

Notice of Intent (NOI)

The NOI describes the LRP and the LRP (entity) stakeholder relationship to the project, as well as the general site and project information. The Operator or LRP will file a SWPPP, describing the means and methods and substantive language of applicable permit compliance implementation that the Operator and his or her workforce will follow during the pre, during, and post construction phase activities of the project.

The NOI and Permit Required Documents (PRDs) for construction activity should have, but are not be limited to, a brief description of the project, estimated schedule and timetables for controlling activities from start to finish, the approximate Disturbed Soil Areas (DSAs) with corresponding revised universal soil loss equation (RUSLE) calculations to determine risk factors for sediment migration and soil loss due to storm water events affecting the site, and ultimately a professionally developed and designed SWPPP.

At the end of the project, a Notice of Termination (NOT) will be submitted to the permitting agency by the Operator or LRP, substantiating that final site stabilization has been achieved per the permit regulations and compliance regarding storm water runoff quality from the project site.

CHAPTER 3

The Storm Water Pollution
Prevention Plan (SWPPP)

Once the determination has been made whether the project is a Phase I or Phase II project, it must be determined if the permit will be for construction or an industrial or multisector general permit for industrial activities. A corresponding SWPPP will be developed according to the type of permit required.

A professionally developed, certified SWPPP plan, prepared by a Qualified SWPPP Developer (QSD in California) is required for permit compliance. The SWPPP is a living, dynamic document. It should be amended as necessary and updated often. It should contain but not be limited to:

1. A description of the scope and intent of the project.
2. Maps showing the location of the project and the affected areas surrounding the project, such as material and equipment storage yards, offices, water bodies, and so on.
3. The SWPPP will name the owner or LRP of the project and the prime contractor who will be implementing and executing the SWPPP.
4. The SWPPP must provide a risk assessment calculation of sediment and soil loss from the site, providing soil types, gradients of affected areas (slopes), historic rainfall data, and so on, using a revised universal soil loss equation (RUSLE) or equivalent. These calculations must be performed by a certified Qualified SWPPP Developer. The RUSLE data, when coupled with proximity to waters of the state, create a risk level, which is the likelihood of a storm water discharge of sediment or deleterious substance to receiving water or a threat to water quality of any type.

Combined Risk Level Matrix				
		Sediment Risk		
		Low	Medium	High
Receiving Water Risk	Low	Level 1	Level 2	
	High	Level 2		Level 3

The above table is from *California Regional Water Quality Control Board Construction General Permit 2009-0009-DWQ amended by 2010-0014-DWQ & 2012-0006-DWQ*, page 29, section I. Background.

Below are the RUSLE compilation factors as found in the revised California Construction General Permit.

Project Sediment Risk:

Project Sediment Risk is determined by multiplying the R, K, and LS factors from the Revised Universal Soil Loss Equation (RUSLE) to obtain an estimate of project-related bare ground soil loss expressed in tons/acre. The RUSLE equation is as follows:

$A = (R)(K)(LS)(C)(P)$

Where: A = the rate of sheet and rill erosion

R = rainfall-runoff erosivity factor

K = soil erodibility factor

LS = length-slope factor

C = cover factor (erosion controls)

P = management operations and support practices (sediment controls)

The C and P factors are given values of 1.0 to simulate bare ground conditions.

There is a map option and a manual calculation option for determining soil loss. For the map option, the R factor for the project is calculated using the online calculator at http://cfpub.epa.gov/npdes/stormwater/LEW/lewCalculator.cfm. The product of K and LS are shown on Figure 1. To determine soil loss in tons per acre, the discharger multiplies the R factor times the value for K times LS from the map.

RUSLE information is taken from Ibid., page 27, section I. Background.

5. Once the risk level calculations are complete, the data tell the QSD which best management practices (BMPs) to implement at which locations and task areas to most effectively eliminate the risk of sediment migration or construction debris leaving the site and affecting water quality. In California, SWPPP projects are Risk Levels 1–3, with Risk Level 1 being the least likely to affect water quality and Risk Level 3 having the most potential to affect receiving water and storm water quality.

6. Risk Level 2 and 3 projects require Rain Event Action Plans (REAPs), which must be completed seventy-two to forty-eight hours before the beginning of predicted rainfall. The REAP is a reporting document that assures the owner of the project that all required storm water best management practices are in place and that requisite means and methods are in place so that there are no threats to water quality due to the forecasted rain event. The REAP may be filled out by a Qualified SWPPP Practitioner (QSP) and should be reviewed by the QSD overseeing the project and the owner of the project.

7. The SWPPP will certify the qualifications of the QSD and the personnel matrix implemented for SWPPP execution, quality assurance and quality control, water monitoring, and so on.

8. The SWPPP will describe which, how, and where: which of the up to fifty-two or more BMPs will be implemented throughout the various stages of the project. Depending on the risk level matrix of the project, if bioassessment monitoring (bioassessment monitoring assesses the effect and potential effect of the project on receiving water living aquatic life, organisms, vertebrates, and invertebrates), active treatment systems (ATS), or water monitoring and sampling are required, these disciplines and their methods of implementation will be described as required in the separate sections of the SWPPP.

9. The SWPPP will have the entire project schedule, with environmental and wildlife work window restrictions and critical path tasking from start to completion. These schedules should be updated regularly as required.

CHAPTER 4

Qualifying Rain Events and Sampling

I want to take a chapter to briefly cover the definition of a qualifying rain event, regular working hours, and when and how often to sample.

In California, according to the Construction General Permit, a "qualifying rain event" is one that produces at least 0.5" of rain. The measurable rainfall must be quantified and qualified by an on-site rain gauge and, if possible, corroborated by a nearby National Oceanic and Atmospheric Association (NOAA) weather data site.

It must be understood that a single rain event, by definition, could last for weeks. A defined event can have periods of up to forty-seven hours and fifty-nine minutes of no measurable rainfall, but have measurable rainfall at the forty-seven-hour and fifty-nine-minute juncture and still be part of the original rain event. As long as there is measurable rainfall without a forty-eight-hour period of no measurable rainfall, it is still the same rain event. Once precipitation begins, only after there is a forty-eight-hour or more time period of no measurable rainfall has that particular rain event ended.

Some sources dictate that there must be at least a tenth of an inch or more every forty-eight hours to qualify as a contiguous rain event. However, the California Construction General Permit states that any measurable rainfall within a forty-eight-hour period constitutes a contiguous rain event.

These amounts of rainfall must be recorded daily, during normal working hours, to quantify the total rainfall for the rain event. If there is rainfall before or after normal working shift hours, the QSD is not responsible for recording and reporting the data.

On a project that operates Monday through Friday from 07:00 to 17:00, no rainfall data is required to be gathered during nonworking hours or on weekends. If rain is forecasted on Wednesday for > 50 percent chance of > 0.5" of rain on Saturday, a Rain Event Action Plan (for Risk Level 2 and 3 projects only in California; it may vary by state) should be submitted by the QSD to the owner of the project at least forty-eight hours prior to when the rainfall is predicted to begin per NOAA or equivalent forecasting.

Storm Water Effluent Runoff Sampling

Once rainfall begins, sampling should begin when and wherever there is storm water discharge from the project site, at each discharge location, which should be documented on the Water Pollution Control Drawings in the SWPPP. If there are ten discharge points on the project, a minimum of one to three samples should be taken at each location, preferably within two hours of the beginning of storm water effluent runoff. A minimum of three storm water effluent samples must be taken per sampling event.

The QSD or his or her designee (who must be trained in sampling techniques by a qualified trainer of record or equal) should always sample whenever storm water effluent runoff begins, whether 0.5" of rain has fallen or not. *Sampling should begin when* storm water effluent *runoff begins*, in case there is only one day of +/- 0.25" rainfall creating runoff and the storm lasts for a week, off and on, creating more than 0.5" of measurable rainfall with only one day of precipitation creating runoff.

The QSD will want to have storm water runoff effluent data for turbidity that should be between 0 and 250 nephelometric turbidity units (NTUs) and pH values between 6.5 and 8.5, depending on the project permit requirements. If the QSD misses the storm water runoff effluent sampling event and has over 0.5" of rainfall, he or she is out of compliance with the SWPPP and Construction General Permit and is subject to scrutiny or ultimately citation or fining.

Another reason why a QSD should sample storm water runoff if the rain event is < 0.5" of rain (if it's < 0.5" of rain, the sampling number reporting values do not have to be reported to the Regional Water Quality Control Board (RWQCB)) is that the pH and NTU readings will tell if the soil stabilization, sediment control, sweeping, and corresponding BMPs are working well enough to get compliant sample Numeric Action Levels (NALs). If the pH and turbidity readings are out of compliance with required NAL criteria, one can adjust or augment the BMPs to implement compliant sampling data values. Numeric Action Levels are numeric values and data that reflect the effectiveness of BMPs, means, and methods of in-place storm water pollution control best management practices to help evaluate and corroborate corrective actions necessary for permitted storm water effluent discharge compliance.

In short, a < 0.5" rain event that creates runoff creates a perfect opportunity for the wise QSD to evaluate the effectiveness of the BMPs in place and make appropriate changes before a qualifying rain event comes that requires storm water effluent discharge sampling and *reporting* to the RWQCB.

And remember, when calculating sampling data, you calculate the *daily average for all* readings. When averaging for turbidity, add all the readings and divide by the number of samples to get your sampling day data. When averaging for pH, you cannot add all the readings and divide by the number of samples, because pH averaging is logarithmic. Go to www.http/wgr-sw.com/pH/ or a similar logarithmic pH calculating website for your daily pH average calculation data if necessary.

Parameter	Test Method	Discharge Type	Min. Detection Limit	Units	Numeric Action Level
pH					lower NAL = 6.5
	Field test with calibrated portable instrument	Risk Level 2		pH units	upper NAL = 8.5
		Risk Level 3	0.2		lower NAL = 6.5 upper NAL = 8.5
Turbidity	EPA 0180.1 and/or field test with calibrated portable instrument	Risk Level 2	1	NTU	250 NTU
		Risk Level 3			250 NTU

Numeric Action Level Table is from California Regional Water Quality Control Board Construction General Permit 2009-0009-DWQ amended by 2010-0014-DWQ & 2012-0006-DWQ, page 28, Section I. Background.

401, 404, and 1600 Permit Sampling

Depending on the permits required on the project—if there is a 401 Water Quality Certification, 404 US Army Corps of Engineers Permit, 1600, 1601, or 1602 Department of Fish and Wildlife Permit, or similar permit—additional receiving water, flow channel, or storm water effluent discharge sampling and monitoring will most likely be required. Be sure to review all specific permit requirements and restrictions prior to developing the SWPPP and Water Pollution Control Plan so that all scheduling and tasks can be congruent with permit criteria.

CHAPTER 5

The Construction Site Monitoring Plan

A SWPPP will always include a Construction Site Monitoring Plan (CSMP). The CSMP describes all monitoring procedures (visual and actual) and instructions, location of forms, checklists, and site maps for the project.

Types of monitoring vary by condition and requirement. For example, at minimum, on most projects there must be a weekly visual walk-through, documented inspection performed by a trained designee. The inspection will be performed by the QSD, a QSP, or a trained designee who has the training, experience, and qualifications to generate a legitimate and accurate project site report.

The report, usually a generic site inspection template, may be handwritten or computer-generated. It is always encouraged to take photos that document compliance and BMP maintenance and installation procedures. The report must be signed and dated to become a valid, executed inspection document.

A daily weather monitoring log is an integral part of executing a successful SWPPP. Most projects use the National Weather Service or NOAA, the National Oceanic and Atmospheric Administration. Depending on the project special provisions, alternate weather sites or data may be used, but the source(s) must be noted and used exclusively for the project. An on-site rain gauge should always be implemented whenever possible. Accurate rainfall data is imperative for proper execution of the SWPPP and weather monitoring data.

Documenting real-time weather data by printing out (or printing electronic PDF files of the real-time weather data) actual and forecasted precipitation and wind speed data is critical for the QSD. If for any reason the SWPPP gets audited by the EPA, RWQCB, or such, it is impossible to go back in time and print out the real-time data that are part of the daily weather monitoring data task as outlined in the CSMP substantive language. I have seen contractors receive $10,000 deductions from their contract bid amount for noncompliance in these issues. The *real-time* weather data cannot be recreated later in the NOAA system.

One of the main reasons for tabulating and collecting real-time storm data is to know when to prepare a Rain Event Action Plan (if applicable) and to know how much rainfall is coming and how it will be deposited, whether over one day or over five days. If a storm has heavy predicted wind events, the QSD will know to take the measures necessary to prevent wind-borne erosion or fugitive solid waste from leaving the site. And if there is a historic rain event that causes significant BMP damage and/ or sediment discharge from the site, depending on the permit, if the rainfall amount exceeds a ten-year or twenty-year storm event rain amount, the owner of the project may have legal recourse if he or she can prove that due to historic rainfall amounts, he or she should not be held liable for effluent discharge from the site.

Pre, during, and post storm inspections are also part of the CSMP. Visual inspection frequency will be detailed in the CSMP language. Visual inspections and monitoring also include documentation of non-storm water discharges, illegal connections and dumping, unauthorized discharges, BMP status, and BMP corrective actions. It is an important part of project and risk management to properly and thoroughly document these conditions or events.

Another part of the CSMP is the Sampling and Analysis Plan (SAP). This plan outlines the requisite sampling and analysis procedures per the SAP triggers. If there are multiple permits requiring receiving water or in-channel work monitoring, those monitoring matrices will be detailed in the SAP. The SAP also includes non-visible pollutant monitoring for pollutants that are not visible in storm water or other discharges. A list of potential non-visible pollutants (NVPs) must be generated by the QSD when the SAP is written, by observing construction means and methods and materials and chemicals used in the project tasks, and by observing existing and preexisting natural site conditions that may convey NVPs in storm water discharges. The SAP for NVPs must include the sampling procedures and schedule for sample collection for applicable permit compliance.

Documentation

Proper record keeping, filing, and documentation are an essential part of successful project contract execution and project management of the CSMP. Performing the inspections, monitoring, training records, and weather data tasks in a timely, congruent manner and in the correct sequence will facilitate project compliance and minimize risk management.

I worked as an independent assurance auditor for five years, auditing SWPPP projects for paperwork and field BMP compliance. The majority of the failing grades of the projects audited were because of improperly performed or missing monitoring reports and training activity documentation as mentioned in this section. Some of these deficiencies resulted in fines to the contractors who were negligent in their monitoring, record keeping, BMP training, and documentation. Most SWPPP records, files, and documentation are required to be kept for at least three years after the Notice of Termination is filed and accepted.

If as an owner or project manager, one thinks that SWPPP is the place to cut compliance, labor, and supervision costs, think again. Those days are over in many parts of the nation and world. In a project bid, nearly all of the SWPPP items are similar in scope and task. One shouldn't lose a bid proposal for a project because of SWPPP bid items. It's a pretty level playing field these days. Just read your plans, addendums, and special provisions so that you can allocate the required funds, labor, and resources for the requisite project compliance.

Many of the task costs related to contract SWPPP implementation are the same on the engineer's estimate for the bidding process. Take SWPPP implementation seriously, and it won't be the thing that negatively affects your project when you least expect it. It's worth it to do it right all the time. It's a good investment in project and risk management, and it helps the owner of the project appreciate project management's attention to detail and professionalism. Contractors and owners who gain a reputation for SWPPP compliance are often more likely to be considered for time and material or change order work offered by state and federal project managers because of their attention to environmental compliance.

Criminal Penalties

The EPA has the right to enforce criminal prosecution on corporations or individuals for Clean Water Act (CWA) violations. Willful and negligent violations of the CWA may be prosecuted by the Department of Justice. These violations may include but not be limited to failure to maintain proper records, failure to install and maintain appropriate BMPs, or willfully and knowingly falsify reports, statements, certifications, and documentation. Knowing and negligent violators may be subject to daily fines from $2,500 to $50,000. Fines for falsifying statements, certifications, or representations can be up to $10,000 and up to two years in prison or both.

An initial conviction for a negligence violation fine can be from $2,500 to $37,500 per day and imprisonment up to one year or both. Known violations when a person is placed in imminent danger of serious bodily injury or death may yield fines up to $250,000 per person, up to $1,000,000 for organizations, and up to fifteen years in prison.

Enter these values into your risk management calculators!

Part Two

Temporary Erosion Control Best Management Practices

CHAPTER 6

Soil Stabilization Best Management Practices (13)

S-1 Scheduling—Scheduling is the only non field implemented BMP, yet it is one of the most important. As aforementioned in this publication, scheduling is definitely one of the most important parts of project management. It shows the tasks and time lines of the entire project and is, therefore, an integral part in planning and execution of the critical path operations required to construct the project in an orderly and cost- and time-effective manner.

The schedule should also incorporate all environmental windows and time lines, as well as SWPPP BMP installation, implementation, and maintenance in phases as the project progresses. The schedule should show when (if) the SWPPP needs to be certified, as some owners of projects require annual certifications of SWPPPs on certain anniversary dates. Wildlife spawning, nesting, and mating windows must be observed as permitted and therefore must be noted in the schedule time lines so that critical path tasks are performed at the appropriate times for biological and environmental compliance.

The schedule should be amended as often as necessary to keep task, phase, and compliance issues congruent with one another. A properly executed schedule, when followed and amended as needed, is an effective project management and communication tool.

SS-2 Preservation of Vegetation or Existing Property—This BMP often includes exclusion fencing, which is high-visibility orange woven fencing constructed of high-density polyethylene material. It is installed by driving metal "T" posts into the soil and affixing the fencing material with plastic zip ties, wire, or an additional specified material approximately every ten feet.

Some exclusion fences are very elaborate, such as frog and snake fences, which are constructed out of plywood and wire mesh to keep the frogs and snakes out of harm's way from construction vehicle traffic.

Oftentimes the fencing is used to delineate environmentally, biologically, or archaeologically sensitive areas from active construction areas and activity.

SS-3 Temporary Hydraulic Mulch (polymer-stabilized fiber matrix)—Temporary hydraulic mulch is a mixture of shredded organic fiber and wood or other pulp containing polymer tackifiers (binders) that, when mixed with water and sprayed or applied on disturbed soil from a hydro seeding applicator spray unit, provides a topical uniform stabilization coating. It is applied wet, but when dry creates a homogenous crust of organic mulch and fiber that stabilizes disturbed soil areas, helping to prevent fugitive sediment migration caused by water and wind-borne erosion forces.

SS-4 Temporary Erosion Control (with temporary seeding)—This uses essentially a hydro seed mixture, similar to hydraulic mulch, containing seed that will germinate under the proper conditions, creating mature ground cover and root growth to help stabilize disturbed soil and protect against storm water flows and raindrop erosion. It is applied with a pressurized hose from a hydraulic mixing tank.

SS-5 Temporary Soil Stabilizer—Acrylic polymers or palliative chemicals that are applied to the soil, causing the soil molecules to bind together, forming a solidifying layer of stabilized soil to prevent water and wind-borne erosion.

SS-6 Temporary Erosion Control (straw mulch with stabilizing emulsion)—Straw that has been hand placed or blown out of a straw blower to help stabilize disturbed soil areas. Proper application includes spraying a stabilizing tackifier emulsion over the straw to bind it together, or straw can be pinned or pushed down into the soil by a mechanical method to create a cohesive blanket of straw. A proper application of straw at a rate of two tons per acre can reduce storm water erosion soil loss by up to 98 percent.

SS-7 Temporary Erosion Control Blanket (also known as Rolled Erosion Control Products or RECP)—There are many kinds and types of erosion control blanket systems, products, and applications. There are applications for slopes, swales, ditches, and all disturbed soil areas. These products are made from jute, sisal twine, coconut fiber, synthetic polypropylene or polyester fabric, and many more materials. Basically the products differ by composition, performance, and durability. Keep in mind that we are speaking of *temporary* soil stabilization products here, not permanent.

Depending on the application required for SWPPP compliance, the proper RECP product is chosen and installed per the manufacturer's plans and specifications for maximum performance. Quite often these BMPs are not properly installed, lapped, keyed in at the top and bottom of the run, stapled properly, or installed per manufacturer's specifications and are therefore not as efficient as they are designed to be. I often refer to the improper installation of these products and similar BMPs as "eye candy." If not installed properly, they look OK to the untrained eye, but are not effective soil stabilization or sediment control BMPs. The manufacturer's specified installation instructions should be followed for maximum effectiveness. Before the RECP blanket is installed, the area must be graded and groomed if possible so that the blanket will have the maximum earthen/soil contact so that there are not tents or raised areas of fabric that are not stapled correctly. The staple grid pattern holding the RECP with downward force maintaining contact with the soil is what makes the RECP fabric effective in deterring raindrop erosion and soil and sediment migration during storm water sheet flow events.

SS-8 Temporary Mulch (wood)—Temporary wood mulch can be a very effective soil stabilization BMP when properly implemented. Oftentimes on projects, when trees and brush are cut down and chipped up for removal, the material is used as temporary wood mulch. The key about the use of wood mulch as an effective soil stabilization BMP is that it must be used in slopes and areas flatter than a 3:1 ratio. If a slope is steeper than a 3:1 ratio, the wood chips tend to migrate during storm water sheet or concentrated flow events and often end up fugitive from the site or in the storm drains or MS4 drainages. If applied properly and generously, wood mulch may be as effective as properly installed straw soil stabilization ground cover.

SS-9 Earth Dikes, Drainage Swales, and Lined Swales—These BMPs can vary tremendously as needed. Earth dikes or berms are often used to manage the flow of "run-on" storm water that enters the site from an up gradient or adjacent area. Storm water effluent flows that may enter the project from off-site should be managed or redirected so that they have minimal or no effect on the project BMP efficiency or implementation.

Drainage or lined swales, also known as bioswales, may be lined with rock, mulch, mature vegetation, or rolled erosion control products to help channel storm water sheet and concentrated flows to minimize on-site soil and sediment migration and implement soil stability to protect against raindrop impact erosion. Oftentimes these BMPs are implemented to divert sheet and concentrated flows away from active project disturbed soil areas during the construction process.

SS-10 Outlet Protection—Outlet protection is found at storm drainpipe and conveyance concentrated flow discharge areas. It is often implemented by the use of rock slope protection (RSP) or riprap, which is rock of variable and appropriate size and gradation placed at the discharge point on top of soil stabilization fabric to minimize discharge point erosion and act as an energy dissipation system to slow the flows of storm water effluent at the point of discharge. Outlet protection may also be rolled erosion control products or Portland Cement Concrete (PCC) slabs or weirs.

SS-11 Slope Drains—Slope drains capture and convey storm water effluent on sloped areas so that sheet or concentrated storm water effluent flows do not effect rill or gulley slope erosion. Slope drains are usually implemented by piping or lined ditches.

SS-12 Stream Bank Stabilization—Stream bank stabilization may be implemented by the use of rock, k-rail, or RECP installation. Depending on the need and application, stream bank stabilization may simply be the installation of jute mesh blankets or other soil stabilization fabric on exposed stream banks. Other applications may call for the installation of rock slope protection and reinforcing fabric and/or the coincidental use of concrete k-rail or similar BMP that will prohibit scouring of the stream bank by storm water flows. Occasionally native indigenous plants or trees are planted in riparian areas as temporary slope protection BMPs, but this practice is usually part of the permanent erosion control plan.

SS-13 Polyacrylamide (PAM)—Polyacrylamide is a nontoxic polymer soil stabilizer, binder, and soil conditioner that augments the properties of disturbed soil to facilitate aeration and porosity, minimizing soil and sediment migration during storm water runoff events. Polyacrylamide also promotes existing plant and root health while decreasing water requirements. It is usually applied by spray application from a hydro seeding sprayer unit.

Polyacrylamide may also be used as a flocculent, added in solution to an active treatment system or sediment basin to floc and facilitate suspended sediment concentration (SSC) settlement in holding ponds or water treatment tanks. PAM flocculates suspended sediment or solids suspended in liquid into heavier solids or molecules so that they will settle out, reducing turbidity NTUs and SSC discharges.

CHAPTER 7

Sediment Control Best
Management Practices (11)

S C-1 **Temporary Silt Fence**—A silt fence is one of the most used and one of the most improperly installed and maintained BMPs on many project sites. Of the miles of silt fence that I have inspected and seen installed, the majority has not been installed or maintained correctly.

Silt fence usually comes in 100' rolls with 2" x 2" x 4' stakes with 36"-wide polyester or polypropylene fabric stapled to the stakes every five to six feet. (These materials and dimensions vary state by state.)

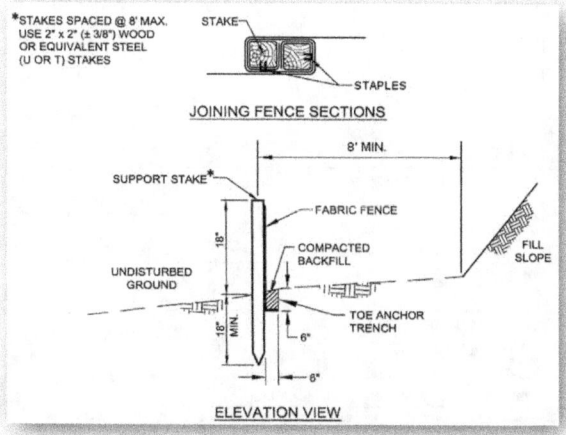

This image is from http://www.elibrary.dep.state.pa.us.

The above image gives a rough idea of silt fence installation techniques. The California DOT (Caltrans) diagram shows the wooden stakes being 48" long, approximately five to six feet apart. The main thing to remember is to trench (key in) and place the bottom of the fabric a minimum of 6" into the soil, with the stakes always being on the downhill side of the fabric, so that storm water or sediment flows push the fabric *into* the stakes of the silt fence, not *away* from them.

There are variations of silt fence, such as regular silt fence referenced above, and reinforced silt fence (see below).

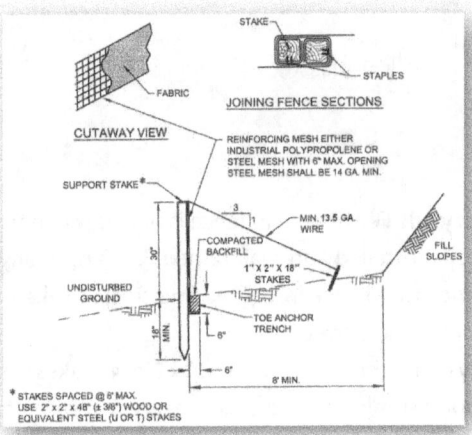

This image is from http://www.elibrary.dep.state.pa.us.

Silt Fence Fabric			
Property	Test	Value	
		Woven	Nonwoven
Grab breaking load, 1-inch grip, lb min, in each direction	ASTM D 4632	120	120
Apparent elongation, percent min, in each direction	ASTM D 4632	15	50
Water flow rate, gal per minute/sq ft min and max average roll value	ASTM D 4491	10–100	100–150
Permittivity, sec^{-1} min	ASTM D 4491	0.1	1.1
Apparent opening size, inches max average roll value	ASTM D 4751	0.023	0.023
Ultraviolet resistance, percent min retained grab breaking load, 500 hours	ASTM D 4355	70	70

The table above is from *California Department of Transportation 2010 Standard Specifications*, Section 88—1.02E, page 960.

Some types of reinforced silt fence use metal "T" posts up to 8' long. Reinforced fences are used when there is a higher probability of rocks, soil, and sediment rolling down steep slopes that may enter MS4 or other drainages if not deterred.

Silt fence should never be installed in concentrated flow areas, such as creeks, gullies, drainage swales, or areas where the fence will be inundated by high storm water flows and, subsequently, fail or be pushed over or bypassed underneath, causing fugitive soil and sediment migration. The silt fence should be properly lapped (wrapped) at each contiguous joint and should always be perpendicular to the slope of down gradient flow of storm water.

Whenever silt fence builds up soil, sediment, or debris greater than one-third of the actual fabric height from the original ground, the deposited material should be removed and properly disposed of. Don't shovel the loose debris or soil down gradient of (behind) the silt fence. I have seen projects where the sediment built up in front of the silt fence was actually shoveled into active drainage conveyances behind the silt fence, defeating the purpose of the silt fence as a sediment control BMP. Maintain silt fence regularly, properly, and wisely.

SC-2 Temporary Sediment Basin—Sediment basins or traps are built or placed in the path of sediment-laden flows or have collection conveyances that convey the sediment-laden effluent to the trap or basin. Sediment traps are usually smaller than basins and typically serve drainage areas of five acres or less, depending on the soil type and sediment molecule size, and whether there are primary, secondary, and tertiary tributaries that convey large amounts of storm water flows toward the settlement basin or trap.

Temporary sediment basins may be in place for several years and are typically designed to handle a minimum ten-year, twenty-four-hour storm water flow event. Depending on the RUSLE universal soil loss equation data, a typical sediment basin sediment storage volume is around 3,600 cubic feet for each acre of disturbed soil that the basin is mitigating. This is the equivalent of one inch of runoff from the disturbed soil drainage area.

Maintenance should be performed before the basin or trap reaches 50 percent sediment storage capacity or less. If the project area doesn't have one area large enough for a single trap or basin, a series of smaller traps or basins can be created with overflow pipes into the next basin, creating additional settlement areas for suspended solids to settle out of the effluent prior to discharge from the basin or trap.

SC-3 Temporary Sediment Trap—A sediment trap can be any appurtenance, human-made or otherwise, that allows turbid effluent-suspended solids to be removed or captured or settle out of storm water effluent. A sediment trap may be a hole dug in the ground, a natural swale that traps water, or a Baker tank or human-made holding tank made of wood, metal, or an impermeable liner in a hole dug in the ground.

The sediment trap may or may not have an outlet. Some simply hold effluent until the water infiltrates into the soil, leaving the solid and particulate matter in the trap. Some traps may have an overflow pipe or drainage appurtenance, which is often filtered by using filtration fabric, clean drain rock, or mature vegetation. (See also SC-2 Sediment Basin above.)

SC-4 Temporary Check Dam—Temporary check dams may be created from any number of methods or materials, including but not limited to drain rock, drain rock wrapped in filter fabric, fiber rolls, native soil, straw bales, rubber or polyethylene, rock bags or rock-filled socks, random rocks, logs, or trees.

For the purpose of the installation of contemporary conventional SWPPP BMPs, normally the check dam materials are purchased at storm water temporary BMP supply houses or rock suppliers. The use of indigenous logs, rocks, and materials are not commonplace in the storm water management plan industry for temporary BMP implementation.

Check dams should be placed strategically on sloped storm water flow areas, perpendicular to storm water flows so that they create steps or barriers, causing the storm water runoff to eddy or be slowed down, implementing silt and sediment deposition and settlement before flowing over the check dam.

Check dams should be spaced, or placed, so that the *top* of the check dam that is downhill from the previous check dam (uphill one) is level with the bottom of the uphill check dam. For example, if check dam "A" is built of rock bags 12" high, the top of the next check dam (check dam "B") downhill should be level with the bottom of check dam "A." Therefore, the proper spacing of the check dams will be determined by the height of the check dam and the gradient of the slope (D below) where the check dams are installed.

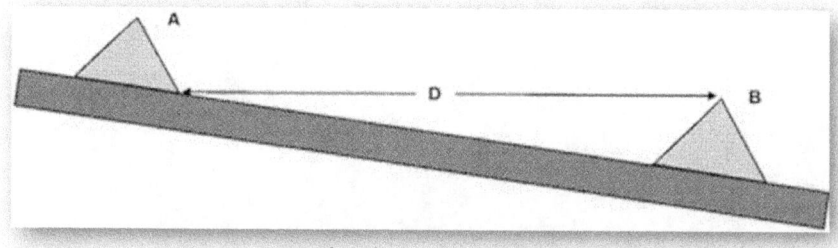

Image from http:www.imgkid.com.

As with silt fences, often check dams are installed and maintained incorrectly, minimizing efficiency and accounting for not much more than "eye candy." Conventional check dam materials often have manufacturer installation instructions and recommendations. But there is no shortcut for having the knowledge and expertise of knowing which check dam materials and methods to use, when, and where. Only an experienced and certified erosion control specialist can take the guesswork out of maximizing check dam installation and efficiency and minimizing silt and sediment migration.

SC-5 Temporary Fiber Rolls—Below are the details for installing fiber rolls per the California DOT (Caltrans) Standard Plans. Having observed other installation drawing details, these are as definitive as any, and the fiber rolls work well when they are installed per these details.

There are two basic types of fiber roll installations: Type 1 and Type 2, per the illustrations below. Type 1 is generally used on flatter slope areas (3:1 or flatter) when the fiber rolls are not installed over excelsior roll or rolled erosion control products on disturbed soil areas/slopes. Type 1 fiber roll installation calls for a 2" to 4" keyway being trenched for the fiber roll to lie in before being staked (see Type 1 installation detail below). Type 2 fiber roll installation calls for an alternated staking pattern on each side of the fiber roll with notched stakes. Sisal twine or its equal is placed in the notches before the wooden stakes are driven to tip depth, creating down pressure on the fiber roll, causing contact with the soil, creating a pervious dam for storm water flows to be filtered through.

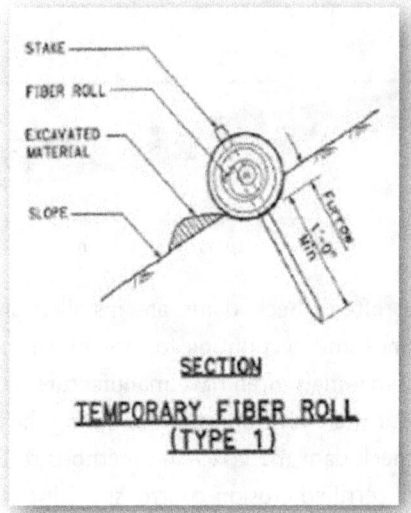

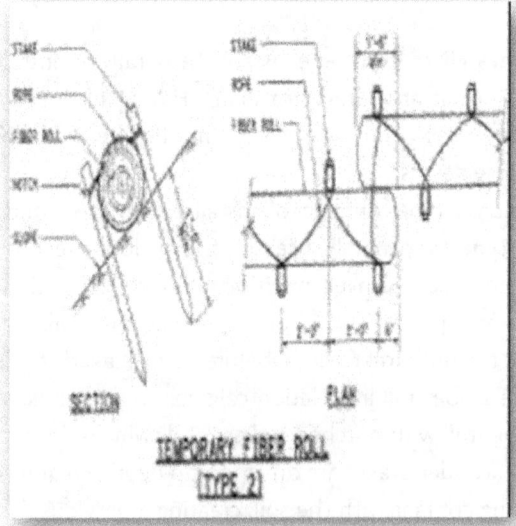

Drawings are from page T56 of *California Department of Transportation 2010 Standard Plans.*

Type 1 fiber rolls are keyed or trenched in, and Type 2 fiber roll installation uses the down pressure of the notched stakes and twine to form earthen contact and form a filtration barrier for storm water sheet flow activity.

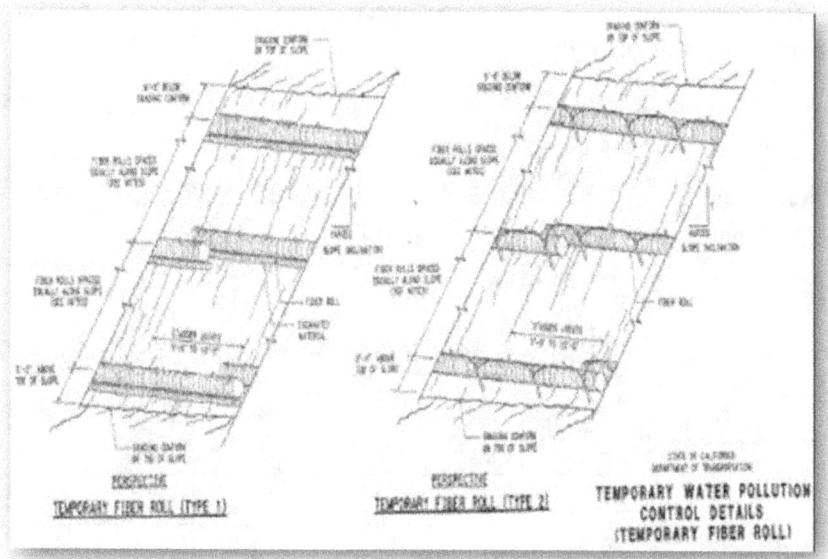

Drawings are from page T56 of *California Department of Transportation 2010 Standard Plans.*

Type 2 fiber roll installations can be used in conjunction with rolled erosion control products on slopes and disturbed soil areas. Proper installation, stake spacing, overlap, and slope orientation are critical for effective BMP performance. Fiber rolls installed on slopes should be installed with a laser or builder's level to ensure that they are perpendicular, or in level plane with relation to storm water sheet flows. If fiber rolls are not installed in a level plane, they will act as water bars, or ditches, conveying storm water flows along the top of the fiber roll, facilitating rill and gully erosion parallel with the slope, instead of forcing the sheet flows to be filtered through the fiber rolls, perpendicular to the slope, per their intended use and function as an effective sediment control and slope stabilization BMP.

It should be noted that fiber rolls, or equal slope protection, should be implemented differently on Risk Level 2 and Risk Level 3 projects, which are projects that have a greater potential to discharge sediment into receiving waters or MS4 drainage conveyances. If your project is a Water Pollution Control Plan (not a SWPPP) disturbing less than an acre of soil or a Risk Level 1 project, these practices are probably not required by the permit.

The following information is taken directly from the California Construction General Permit amended edition:

Additional Risk Level 2 Requirement: Risk Level 2 dischargers shall apply linear sediment controls along the toe of the slope, face of the slope, and at the grade breaks of exposed slopes to comply with sheet flow lengths[3] in accordance with Table 1.

Table 1 - Critical Slope/Sheet Flow Length Combinations

Slope Percentage	Sheet flow length not to exceed
0-25%	20 feet
25-50%	15 feet
Over 50%	10 feet

Per pages 5 and 6 of Attachment D of *California Regional Water Quality Control Board Construction General Permit-2009-0009-DWQ amended by 2010-0014-DWQ & 2012-0006-DWQ.*

This table and information apply to Risk Level 2 and 3 projects only. If you are in a state or region that determines receiving water risk levels similar to California, you may want to consider implementing these practices. Essentially, if your project has slopes of 4:1 incline or flatter, place your linear slope protection every twenty feet vertically. If your slopes are inclined between 4:1 and 2:1, space your linear protection every fifteen feet or less vertically. If your slopes are steeper than 2:1, place linear sediment barriers every ten feet or less vertically.

SC-6 Temporary Gravel Bag Berm—Gravel bag berms are effective filtration and delineation BMPs when properly implemented. A gravel bag berm should be constructed out of a polypropylene, polyester, or geosynthetic mesh bag, 24" to 32" long and 16" to 20" wide, and filled with 35 to 50 pounds of washed rock, 3/8" to 3/4" is most common.

Gravel-Filled Bag		
Property	Test	Value
Grab breaking load, lb, 1-inch grip min, in each direction	ASTM D 4632	205
Water flow rate, gal per minute/sq ft min and max average roll value	ASTM D 4491	80–150
Permittivity, sec^{-1} min	ASTM D 4491	0.2
Apparent opening size, inches max average roll value	ASTM D 4751	0.016
Ultraviolet resistance, percent min retained grab breaking load, 500 hours	ASTM D 4355	70

Above chart is from *California Department of Transportation 2010 Standard Specifications*, section 88–1.02G, page 961.

The proper gravel-filled bag material is of the utmost importance for a number of reasons. The material must be ultraviolet resistant and have enough tensile strength to withstand foot and vehicle traffic and maintenance activity. Often I have seen the inexpensive white, woven-tape fabric bags from the local discount hardware store used for rock or sandbag implementation. These white bags are fragile, are not ultraviolet resistant, and more often than not end up as solid waste with their contents spilled out on the ground.

We must recognize also that there is a direct difference for *sandbag* berm application, as opposed to gravel-bag berm implementation. At times on a project, an actual impervious berm (barrier) is needed to redirect run-on storm water flows or on-site flows, similar to an asphalt curb or dike. I have actually seen gravel bags filled with sand or dirt for this application, but remember, when water hits the bags, whatever is in the bags will leach out of the wall of the bag and into storm water flows. So, please consider how you construct a bagged berm for storm water BMP purposes. I have actually encapusaled rock bags in 6 to 10 mil plastic, forming an impervious "burrito" that worked better than using sand or dirt bags from a storm water runoff/water quality perspective.

When filling gravel bags, be sure to use clean, washed rock. If gravel bags are filled with dirty rock or soil enters the bags when they're being made up, the sediment particulates will leach out of the gravel bags during storm water flows. This can affect turbidity and even pH readings in concentrated flow areas. If the gravel comes from a limestone quarry or a similar high-pH rock source, the calcium, carbonate, or alkali molecules can leach out of the rock and into storm water flows, raising pH numbers to unacceptable levels. (pH readings should be between 6.5 and 8.5 NTU.)

Placement and configuration of the gravel bags are important. Be sure that bags are lapped on top of one another by about one-quarter to one-third of the length of the bag. There should be no gaps or holidays under the bags, and they should have positive earth-to-bag contact. Each bag end should be securely closed so that no gravel can escape.

SC-7 Street Sweeping—Street sweeping should no longer be performed with a kick broom power-driven sweeper. Any sediment control sweeping activity should be performed by a commercial-grade vacuum truck sweeper with multiple rotating bristles and water spray nozzles to minimize fugitive dust and particulates from the street-sweeping activity. Kick brooms spray dust, dirt, and particulates over the entire work area, creating a plethora of storm water management plan sediment control issues.

So many times I have seen an inadequate or antiquated sweeper truck pushing dirt around in a cloud of dust, while in reality, this type of activity is breaking storm water and air pollution control laws, and is a substandard, out-of-compliance best management practice.

Street sweeping is a last line of defense to keep soil and sediment from leaving the disturbed soil areas of the project site and entering the traveled way to off-site locations. The main objective of street sweeping and sediment control practices is to keep off-site tracking of soil and sediment from occurring. Part of the problem of sediment buildup on roadways is that, when it rains, the sediment particulates that have entered the voids in the pavement on the traveled way areas of the project are conveyed from the roadways to any existing drainages or discharge points.

Since storm water effluent monitoring should occur during the first two hours of storm water effluent runoff, turbidity readings may very well be out of compliance, simply because of the buildup of sediment particulates in the paved traveled ways of the project being washed into effluent discharge drainage areas, where storm water sampling must be performed.

Make sure that disturbed soil construction entry/exit points are adequately protected to minimize the need and frequency of street sweeping. Remember, a clean roadway doesn't have to be swept. Contractors and owners alike often underestimate the importance of proper construction entry/exit area BMPs. So to lower the street-sweeping budget, spend the preventative maintenance funds on stopping or deterring tracking from entering into the traveled way.

Street Sweeping Risk Management

Last but not least, read your contract and know where to store and dispose of street-sweeper waste. Storm water pollution control, Title 22, and Code of Federal Regulations laws prohibit the comingling of solid waste with hazardous waste. Know the difference between hazardous waste and solid waste. If street sweeping is picking up asphalt grindings, lead-based thermoplastic traffic stripe waste, or other asphalt or concrete waste that has diesel fuel or cure products comingled with the swept-up material, it is a *hazardous material waste* disposal situation, not a *solid waste* disposal situation.

When in doubt, protect and isolate the material in question and have it characterized by a hazmat laboratory for proper disposal. This may be an extreme part of street sweeping as a storm water BMP, but so many times, I have seen street-sweeper waste piles littered with plastic quart motor oil and chemical containers or similar polychlorinated biphenyl (PCB) or volatile organic compound (VOC) containers. Current storm water pollution control laws prohibit these practices. Know what you're sweeping up and where and how you're disposing of it within permit compliance guidelines.

SC-8 Temporary Sandbag Barrier—The temporary sandbag barrier is, of course, different from the SC-6 Temporary Gravel Bag Berm. For a cross comparison, see also the description of the SC-6 Temporary Gravel Bag Berm for bag specifications and additional information regarding the temporary sandbag barrier.

As referenced earlier in the SC-6 gravel bag berm language, the temporary sandbag barrier creates more of an impervious barrier that directs water flows to a desired conveyance location or keeps storm water effluent flows out of areas where erosion can be started or facilitated. Sandbag barriers should be constructed of the aforementioned compliant bags and should be filled with clean, washed sand, so that if storm water flows come in contact with the bags, sediment will not leach out of the bags, contributing to potential additional suspended sediment concentrations in storm water effluent runoff.

Sandbag barriers may also be wrapped in plastic or polyethylene fabric to form a more durable and impermeable barrier to impede hydraulic migration into areas where effluent flows must not enter. The bags should be overlapped as needed to maintain an impermeable barrier. Each bag should be securely closed so that the sand cannot leach out and become potential fugitive sediment.

SC-9 Temporary Straw Bale Barrier—Straw bale barriers are more commonly implemented in swales or concentrated sheet flow areas where rock bag check dams or silt fence cannot handle increased flows during storm water rain events. Oftentimes the straw bales are secured with rebar or steel "T" posts, anchoring them into the soil.

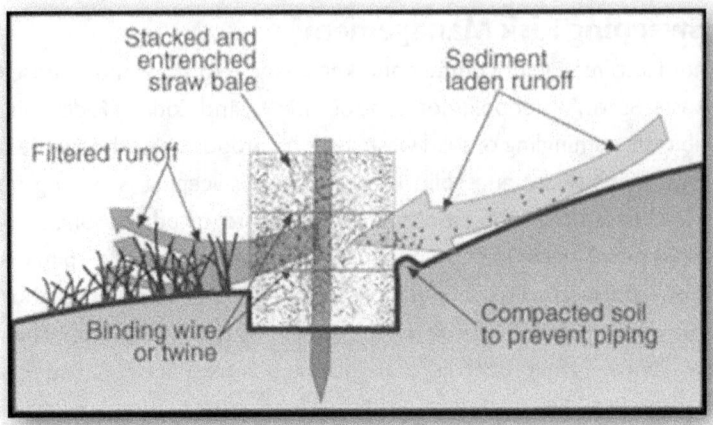

Image from http://www.google.com/urFextension.missouri.edu.

Straw bale barriers should not be used in concentrated flow areas because they cannot filter the water fast enough and often the bale barriers become inundated by large volume flows, resulting in bale berm failure or storm water effluent bypass around the BMP. Straw bale barriers can be implemented with or in between other BMPs such as check dams as effective sediment control/filtration BMPs. Filtration fabric may be placed along straw bale barriers for additional filtration performance.

Be sure to check your project special provisions and permits, because sometimes the straw or bale material has to be certified to be free of certain seed or vegetative matter, especially in coastal or environmentally sensitive areas. You cannot introduce nonindigenous plant seed into some project areas. This is a valuable consideration to be aware of, because once you are done with the straw bale barrier, it may be advantageous to be able to spread the straw material on the ground for additional temporary or permanent erosion control protection, instead of removing it from the site as solid or vegetated waste.

SC-10 Temporary Drain Inlet Protection—There are many forms and types of drain inlet protection, from the very sophisticated to the very basic. Sophisticated temporary drain inlet protection can have oil/water separators, as well as sediment bag inserts that capture solid and/or petroleum-based waste and large particulate matter that enters the drain inlet (DI).

The main objective of DI protection is obviously to not allow silt, sediment, solid, or hazardous waste to enter a drain inlet, whether active or inactive. Oftentimes during construction projects, drain inlets are rendered inactive due to the construction process. If these drains are not covered or protected, sediment or waste may enter the drainage system, only to be flushed into an active water of the state or municipal separate storm sewer system (MS4) conveyance after the DI is reactivated. All drain inlets on an active construction project should be protected in one way or another if they will be affected in any way by the construction of the project.

Another consideration for the type of DI protection to use is to be aware of the approximate flow rate of storm water effluent flows entering the DI and what is up gradient of the DI. Are there disturbed soil areas or factors that will contribute to silt or sediment delivery to the DI during the construction process or rain events? Perhaps a series of check dams are required up gradient of the DI so that silt or suspended solids can settle out before reaching the DI, which will be a sampling point during storm water effluent discharges.

Drain Inlet Protection Challenges

Sometimes drain inlets are in central locations that receive a lot of storm water run-off, but the DI still needs to be protected. Drain inlet BMPs can restrict storm water runoff flows and cause ponding and flooding along traveled ways, which are safety issues. Many times I have seen filter fabric used as DI protection over the inlet grate. The filter fabric gets inundated with silt and clogs up, not allowing storm water runoff to enter the DI, resulting in ponding or flooding of the area. Then during a storm, the maintenance crew cuts a hole in the fabric or removes it from the grate, dumping the ponded-up storm water effluent unfiltered into the drain inlet.

Try to think these issues through *before* a big runoff event and anticipate effective alternatives for compliant DI protection before getting caught in the middle of a storm and having to scuttle the DI protection and allow sediment-laden storm water effluent runoff to enter an MS4 drain inlet or waters of the state. While safety always takes top priority for all activities, having to uncover a clogged drain inlet to relieve storm water ponding along a public traveled way may be avoided while at the same time maintaining storm water pollution prevention compliance if the proper BMPs are implemented.

Drain inlet precast boxes must also be protected during the construction phase of drainage systems. I have seen multiple-piece precast box sections partially installed without the soft asphalt gasket material placed as the watertight seal between the precast sections where storm water has built up in the hole around the precast DI sections and infiltrated the drainage outlet pipe, discharging sediment into receiving waters.

It doesn't look good when a Regional Water Quality Control Board inspector finds fresh silt and sediment discharged into a receiving water of the state because the drain inlet sections were not protected from silt incursion during a storm water rain event. Protect drain inlets during all phases of construction or be prepared to take the risk of discharging deleterious materials or sediment into MS4 drainages or receiving waters. Risk management tells you to protect your drain inlets at all times or risk being out of permit compliance.

If there is a risk of having petroleum-based PCBs or VOCs entering drain inlet areas, hydrocarbon booms may be used to absorb petroleum-based chemicals in areas like parking lots or paved areas where hydrocarbon-based materials have built up over time.

Drain inlet hydrocarbon skimmer found at http://www.google.com/www.biocleanenvironmental.com.

SC-11 Temporary Chemical Treatment—The primary objective of temporary chemical treatment (TCT) is to target sediment mitigation or removal from storm water effluent that is either being held in holding ponds or tanks or treated as a batch system. Turbid storm water that has a high NTU concentration, generally greater than 250 NTU, due to suspended sediment concentration may require TCT to cause the suspended solids to settle out prior to discharging the storm water effluent, where numeric turbidity standards must be met.

Synthetic neutral polymer chemicals such as polyacrylamides or polyethylene oxide are suitable for use as flocculants and coagulants to help sediment molecules bind to one another and settle out or become large enough to be filtered out through sand media filters or equal filtration devices. For clarification, coagulants, such as polyacrylamides, are chemical agents that cause clay or sediment platelet particulates to attract one another and form flocs, which are an aggregation of organic particulate molecules. The flocs are formed by chemical interaction with the coagulants. Basically, coagulation neutralizes or destabilizes the electrostatic charges of particles, and flocculation refers to the agglomeration of the destabilized particles into a flocculent particulate mass.

Since a 3 micron (0.003 of a millimeter) particle of silt or clay may remain in suspension for over seventy-two hours, coagulant polymers are introduced to break the zeta potential, or repelling action, of the micro particulates, causing them to attract one another, forming flocs, and greatly decrease the suspended sediment concentration time. The flocs will then settle into the sediment basin or holding tank or be filtered out by a sand media pump and filtration process via an active treatment system. Be sure to monitor the effluent discharge for residual chemicals, if any, to ensure compliance with allowable permitted discharge levels for pH, turbidity, and any other elements or chemicals of concern. Common effluent standards are 6.5 to 8.5 for pH and 5 NTU for turbidity. If biomonitoring is required for effluent discharge, acute toxicity testing may be required per permit requirements. Know your permit.

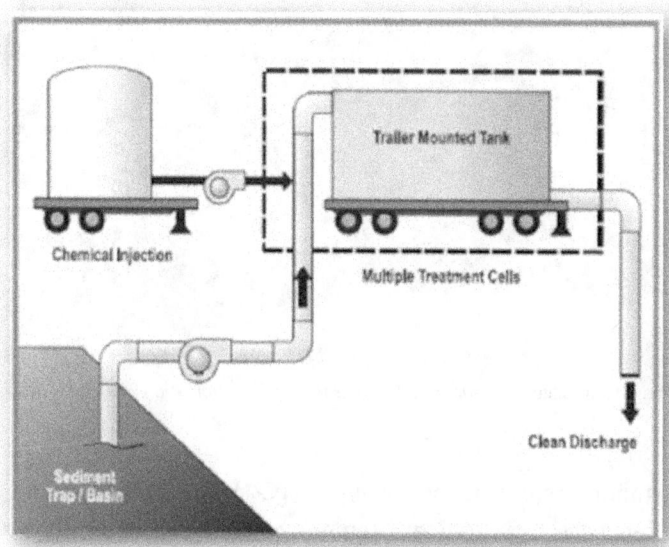

Illustration taken from November 2011—Page. SE-59—City and County of Honolulu Construction BMP, City and County of Honolulu, Hawaii.

CHAPTER 8

Tracking Control Best Management Practices (3)

TC-1 Temporary Construction Entry/Exit—

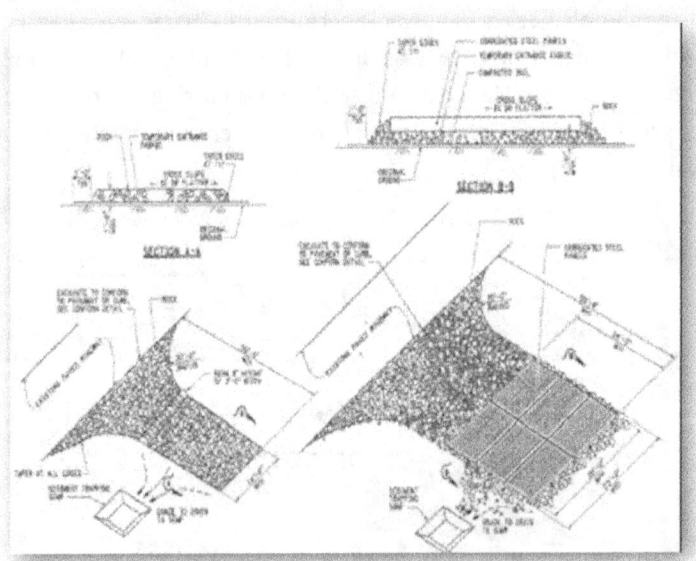

Drawings from page 253, *California Department of Transportation 2010 Standard Plans.*

The temporary construction entry/exit (TC-1) drawings above are a good baseline reference for a standard temporary construction entry/exit. Basically, temporary construction entries/exits should be constructed by excavating the location 1' deep, at least 25' wide and 50' long. After the excavation, if possible or necessary, provide a drainage conveyance (pipe or rock-filled trench or equal) that will facilitate drainage or percolation of built-up storm water from the TC-1 during storm water rain events.

A layer of synthetic reinforcing stabilization fabric should be placed on the ground before placing the 1' of rock. The rock should be clean, not mixed with soil or delete-rious debris. Depending on the project specifications or special provisions, the rock size is normally 3" to 6" minus rock, which meets project-specified durability specifi-cations. The rock must be crushed rock, not cobbles, or it will displace and migrate. Some projects that have a lot of heavy truck and equipment traffic with dual wheels sometimes specify smaller rock so that the rock won't get caught in the dual wheels and be deposited in the traveled way, creating a traveled way obstruction and safety hazard after leaving the TC-1. This should be a consideration, depending on the safest means and methods to be implemented while not compromising the effectiveness of the temporary exit/entrance for storm water management plan implementation.

The main function of the temporary construction entry/exit is to provide a long and rough-enough uneven surface for construction equipment to travel on before leaving the active construction area, which will remove and dislodge mud, silt, or sediment from the equipment tires or tracks. The TC-1, if properly located and constructed, should all but eliminate, or at least control, the amount of sediment that is tracked off the active construction area onto public roadways or traveled ways. And per the aforementioned SC-7 street-sweeping BMP, effective street sweeping should be done coincidentally as needed at all TC-1 locations.

Most TC-1 BMPs that I have inspected over the years are either incorrectly constructed or improperly maintained, resulting in off-site tracking and tracking into public traveled ways. Soil and sediment deposited onto asphalt or concrete paving will become slippery during rain events or being accidentally watered by dust control or other construction activities. And if the affected area of the roadway becomes wet, the soil gets tracked even farther off location and potentially into MS4 or other drainage conveyance areas, causing potential high-turbidity readings for storm water effluent monitoring. Sometimes steel-finned corrugated rumble plates are used, either by themselves or with the rock, to augment the effectiveness of the TC-1 implementation to remove sediment and mud from tires or tracks.

Another inherent problem with TC-1 implementation is that temporary construction entries/exits are not constructed per plan. Often they are too small and basically ineffective. The contractor gets paid "each" as a lump-sum SWPPP bid item, so often they are constructed with minimal effort, materials, and attention to functional detail. Rather than performing marginally in constructing the TC-1, why don't we do it correct the first time? Why? Because if it's constructed *per plan* and is not working as effectively as desired, approach the owner's resident engineer or project management personnel and ask for additional funds to perform a field adjustment, addendum, or redesign to cure the deficiency. Most often, the owners of the project will work with the contractor and share the cost when weighed against the risk management costs of accidents, safety violations, or storm water pollution violations or citations that could affect both parties. And remember, the onus of culpability falls upon the entity that is most out of compliance with the project permits, plans, and specifications. So build it right the first time. Storm water management plan compliance and effectiveness should weigh heavily in the balance of risk management cost effectiveness.

Temporary construction entries/exits should also be constructed and oriented so that construction traffic cannot bypass the BMP. It's there for a reason. Use it or fix it so that it becomes what it's designed to be, a sediment deterrent for off-site tracking.

Often throughout the course of a long construction project, TC-1s are added, omitted, or modified. This is normal. Be looking for the best way to implement a TC-1. If the work is in the median of an interstate highway behind the concrete k-rail, the TC-1 design may need to be reconsidered, or just use corrugated steel plates or another effective measure. Put public safety and storm water compliance in the same boat. They are the motor and rudder. You need them both to get where you're going for compliance. Off-site tracking at a location like this can be dangerous and out of compliance for both safety and storm water pollution.

TC-1 Maintenance

Most temporary construction entries/exits are not maintained properly, if at all. Remember, BMP *maintenance* is part of the storm water management plan. There are weekly inspection forms that document deficiencies and deficiency corrective action maintenance, called corrective action plans. BMP maintenance is part of the bid item for all SWPPP temporary BMP implementation items. If maintenance becomes excessive or ineffective, have the QSD and project management personnel work out an amendment or amicable solution. Don't just ignore the problem until it becomes an insurmountable one. Partnering in situations like this goes a long way for both parties and can have a very positive overall effect on the project.

TC-2 Stabilized Construction Roadway—A temporary stabilized construction roadway (TC-2) is implemented for various reasons, such as construction right-of-way that is soft, wet, or not suitable for heavy construction traffic. Or it may be a haul road that must be stabilized for when earthmoving or construction operations cross a public traveled way, so that mud, silt, or sediment will not be tracked from the construction-disturbed soil areas onto the public traveled way. TC-2s differ from TC-1s in the fact that they are usually longer and may be constructed of smaller durable 3/4" to 2" rock. I have seen clean base rock used as well. Check the project special provisions and standard specifications for the specified compliant rock and stabilization fabric materials. Most DOT specifications call out 3" to 6" minus crushed rock or #25 railroad ballast, which is 3" minus with some fines in it.

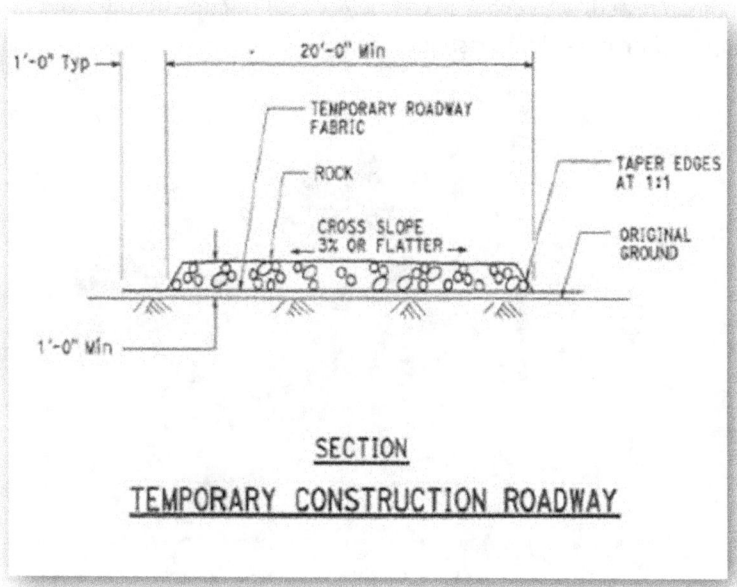

Drawing from *California Department of Transportation 2010 Standard Plans*, page 262.

Be sure to read your special provisions before purchasing and/or placing the TC-2 fabric and rock if you are in a creek bed, riparian zone, or environmentally sensitive area or if the TC-2 drains to or is in a floodplain of a riparian zone. The Department of Fish and Wildlife Service and other environmental permits have certain rock and fabric specifications on projects that are located in riparian zones. And depending on the project, some TC-2s have to be removed by certain work window restriction dates for permit compliance.

TC-3 Temporary Entrance/Outlet Tire Wash—Temporary entrance/outlet tire washes (TC-3s) are high-pressure wash systems that have metal steel racks that equipment vehicles drive over, where high-pressure sprayers wash and rinse off the underbelly and tires of the vehicles. The mud that falls off the vehicles is collected and stockpiled for hauling off to a proper dumpsite (the material should be characterized prior to off-haul to ensure that PCB, VOC, or similar contaminants have entered the effluent material) while the effluent water is captured below the racks and pumped into holding or filtration tanks. The effluent soil may have a designated area on the project site where it may be placed after drying out, similar to burying aerially deposited lead (ADL) soil, where if it does not have harmful contaminants, it can be used as fill material. If the effluent solids do contain hazardous chemicals, then they must be protected from wind and rain events prior to being manifested, with chain of custody (COC) documentation, and off-hauled to an appropriate hazardous waste dumpsite.

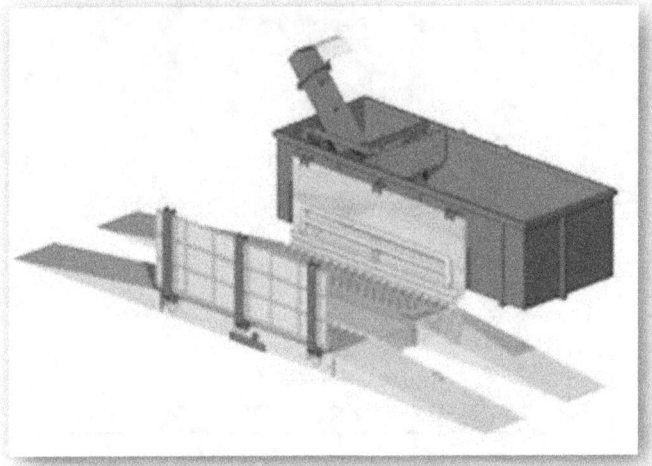

Temporary Tire Wash BMP from www.rainforent.com.

The wash/rinse water may contain polymer flocculants or coagulants to aid in the settling out of solids and particulates, which can be filtered or removed from the system. Tire washes can be expensive to set up and maintain. They are primarily used where space is limited for construction vehicle traffic to have mud and silt removed from tires and tracks before entering or crossing a public, off-site traveled way.

The setup and implementation of a TC-3 must be looked at very closely from a cost and risk management perspective. If at all possible, obviously the tire wash BMP should be avoided due to cost of setting up and maintaining it, but if it must be implemented because of cost-effectiveness over alternate methods or storm water management plan BMP cost and effectiveness vs. permit noncompliance, keep the permit compliance on the heavy side of the balance scale.

CHAPTER 9

Wind Erosion Control Best
Management Practices (1)

WE-1 **Wind Erosion Control**—There are at least three types of wind erosion: suspension, saltation, and surface creep. Suspension is when very fine particulates (< 0.05 mm) become airborne and create fugitive dust. Saltation is the short or intermediate movement of 0.05–0.5 mm particulates by wind events. These particulates do not travel far until they land back on the ground, loosening or dislodging other soil, creating over half of windborne particulates. Surface creep is the movement of particulates of => 0.5 mm along the surface of the ground. Depending on wind speed and volume, these particles often never leave the ground, but migrate, potentially leaving the site as fugitive particulates and dislodging other particulates along the way, compounding the wind erosion process.

Depending on type and size of soil particulates, windborne erosion can begin around 13 mph. A visual inspection of disturbed soil areas and soil stockpiles on windy days can dictate when wind erosion control (WE-1) needs to be implemented. Wind erosion control BMPs for soil include but are not limited to watering, palliatives, hydraulic mulch, and covering with RECPs or plastic. If soil is to be protected with RECP fabric or plastic, be sure to know the proper techniques for anchoring these BMPs well to the affected area, because remember, you're protecting soil from *heavy wind*, so *prepare for heavy winds* and install your BMPs properly. These means and methods will be covered more in detail in WM-3 Stockpile Management.

When using water to keep dust down, remember not to water so much that a non-storm water discharge is created, especially if the dust control is being performed on disturbed soil areas (DSAs) near MS4 drainage conveyances, other drain inlets, or receiving waters of the state. Don't create a storm water management plan nightmare by creating a non-storm water discharge. If a water truck discharges water that ends up entering a drain inlet, receiving water, or MS4 conveyance, it is not reported and documented, and the NPDES permitting entity finds out, the owner of the project and the contractor are culpable and liable for any deleterious effects to water quality or permit requirement noncompliance. Remember, *only rain down the drain*, not water from a water truck or watering system.

However, there are acceptable circumstances when non-storm water discharges are allowed and permissible on projects. Know your project NPDES permit and know what is allowable. If the language of the project-specific permit is nondescript or vague, be transparent and proactive with the permitting agency and get clarification. Additionally, if the owner of the project has an outdated NPDES permit, do your due diligence if you are the LRP or the contractor to make sure that your project tasks, means, and methods are covered, even if by special addendum by the permitting agency.

When using rolled erosion control products, use filter fabric if possible (rather than plastic) to cover DSAs or stockpiles because filter fabric promotes infiltration of storm water into the protected soil area and deters runoff or sheet flows from the protected areas. Plastic or impervious cover can create a runoff situation at an inopportune time or location because of its imperviousness, when allowable infiltration from filter fabric might be a better alternative than allowing or creating the additional runoff.

An additional risk management consideration when implementing hydraulic mulch to cover stockpiled soil to protect against wind-borne erosion is the use of polyacrylamide-based mulch or binder fiber matrix (BFM) tackifier. There are two things to consider here:

1. If a polyacrylamide-based mulch is applied too generously to DSAs or stockpiles in areas where runoff from the protected area flows to storm water effluent sampling points, the polyacrylamide polymers may raise the pH readings above the 8.5 upper numeric action level limits due to their alkalinity.

2. If the soil being protected by the mulch BMP will be used for building a structural fill on the project where compaction testing is required, if too much hydraulic mulch or equal organic product is applied to the protected area, it could affect the compactability or quality assurance density testing in the future. It may also affect the cohesive properties and the plasticity index and nature of the soil where the mulch is applied.

When addressing the wind-borne erosion BMP, we must also address the solid waste issue. (Solid waste management, WM-4, will be addressed later in this publication.) In the storm water pollution world, one of the operative phrases for noncompliance or noting deficiencies is "Did the _ _ _ _ _ leave the site?" meaning, did the sediment, storm water effluent, mud, construction materials, or solid waste get discharged off the project site to private or public lands or waters of the state?

When sediment, soil, effluent from any source, chemicals, construction materials, or solid waste leave the project site due to storm water management plan (SWMP) and SWPPP BMP negligence, then storm water pollution laws and permitted activities have been ignored, compromised, or broken. If broken bags of cement are whisked into a pond by a high-wind event, introducing cement into the waters of the pond, killing the aquatic life in the pond, storm water pollution BMPs and laws have been compromised, and the negligent party is culpable to cure the damage caused by the deficiency/negligence. When thoughtfully and properly addressed and implemented, wind-borne erosion control will effectively remove or diminish high-wind erosion control risk management.

CHAPTER 10

Non-Storm Water Site Management
Best Management Practices (15)

NS-1 Water Control and Conservation—This BMP is to determine if water sources used for project construction are properly implemented according to the SWPPP, project special provisions, and any applicable permit language. If fire hydrants or other public, state, or federal water sources are used for the construction process, there must be evidence in place that the water sources are metered, monitored, and properly compensated if applicable. The water source must not leak, creating a non-storm water discharge. The water tank or holding source must be labeled "Non-Potable Water" if applicable.

Water sources must be approved for use. I was put in charge of a project that had been started about a month prior to my arrival. My first day on the project, just after I showed up, a state park ranger showed up and asked me why the project was drawing water with no permit out of a seasonal creek that had juvenile steelhead fry in it at the time.

After a trip to the ranger station, I returned with a permit and a citation. We were able to draw water from the creek with an emergency permit, but at the end of the project, I had to attend a meeting with the district attorney and the owner of the company to discuss the monetary amount of the fine for drawing non-permitted water from the waters of the state.

From a risk management standpoint, procuring the water use permit prior to starting the project would have been much more cost-effective than paying the fine and citation for being out of permit compliance. Enter that value into your risk management calculator! The equation would look something like this: X (permit compliance) − Y (non permit compliance) = -($)!

NS-2 Dewatering—A book could certainly be written entirely about dewatering risk management. Years ago on construction projects, contractors would dewater holes, excavations, ponding areas, and so on by just pumping the water to another location, which oftentimes could have been a receiving water, environmentally sensitive area, MS4 drainage, or a combination of all.

Today, thanks in large part to NPDES permits, most dewatering effluent discharges are monitored and accounted for per the SWPPP, water pollution control plan, and basin plan permitting documents.

If necessary most projects will have dewatering listed in their task and bid description documents as part of the scope of work before the project ever starts. If dewatering is anticipated as part of the means and methods of the bid items, the owner of the project should see that a dewatering plan is submitted and implemented by the contractor performing the particular task(s) where dewatering will most likely be required.

If you are a contractor bidding on a project where dewatering is anticipated due to the scope of work and the owner of the project does not submit or request a dewatering plan or bid item, it should be a red flag. The problem is that *compliant* dewatering can be very tricky in environmentally sensitive, metropolitan, and highly populated areas if the proper infrastructure or permit language is not accurate or in place to accommodate compliant dewatering.

For example, while inspecting a project in the San Francisco Bay Area, I saw that the contractor was dewatering, or pumping standing storm water effluent into Baker tanks for holding. When asked what the contractor did with the effluent water, they said that they used it in their water truck to water the site for dust control. Observation revealed that the contractor was discharging the storm water effluent into the local publically owned treatment works (POTW) sanitary sewer system, unmetered. This is not only a SWPPP, NPDES violation, but a municipal POTW violation subject to citation and fines to the contractor and the owner of the project.

Another dewatering violation/conundrum that I experienced was the way that a contractor was performing dewatering by pumping heavily silt-laden water from a Cast-in-drilled-hole (CIDH) pile drilling operation into a water truck, then discharging the water around the site, coating the entire site with silt and sediment particulates. The site was within the riparian zone of one of the most viable trout and salmon, cold, spawn, and migratory waters of the state. The water truck was discharging sediment in the riverbed gravel and cobble rocks and the surrounding area. And the oversight resident engineer had not even signed the SWPPP prior to work beginning. This was one of the most egregious projects for SWPPP noncompliance that I have ever seen.

Another problem with dewatering is that if the contractor does not submit a dewatering plan, which often can take up to sixty days or more for approval if multiple permittors are involved, delays to the project may become fatal flaws. For instance, if a contractor is drilling CIDH pile holes in a riparian zone that has wildlife date restrictions, a project may be delayed until the next calendar year if permit window restrictions are not met because no dewatering plan was approved prior to beginning work. Much of the littoral and riparian work under the auspices of the US Army Corps of Engineers and the Department of Fish and Wildlife has designated cold, spawn, and migratory water body work window restrictions.

For risk management purposes, even if a dewatering plan is not in the scope of work and it may be needed to meet critical path scheduling, there should be an agreement in place among the owner, the contractor and the permittor(s) for a contingency dewatering plan.

Quite often in SWPPP language, daily documented inspections must take place when designated activities are performed. Dewatering is one of these activities. Technically, when dewatering occurs, there should be a documented daily inspection of the activities performed and the status of all applicable BMPs.

Basin Plan Dewatering vs. Storm Water Dewatering

We must differentiate basin plan dewatering from storm water dewatering. A basin plan is a standalone water pollution control plan for all water management resources covering a specific geographical region. Basin plan dewatering differs from storm water dewatering because basin plan dewatering deals with storm and non-storm water management of water resources, while storm water dewatering deals exclusively with storm water runoff and effluent discharge. For example, say that there is an active treatment system (ATS) on a project site that deals with treating dewatering effluent from CIDH pile drilling. The ATS system and its effluent discharge criteria would be under the basin plan permitting, not storm water permitting. If a project has sediment basins or an ATS that treats storm water effluent discharge, then storm water pollution permit criteria would be in effect for the sediment basin and ATS effluent discharge data.

Take dewatering seriously and know the parts of the puzzle that may be needed, as well as how to implement the required ones within permit compliance parameters.

NS-3 Paving, Sealing, Saw Cutting, and Grinding Operations—I can remember using ground asphalt and recycled concrete for shoulder backing and roadway construction subbase or base rock material. Probably in some parts of the country, these practices are still used today. But in environmentally sensitive regions where storm water runoff can carry high pH, hydrocarbon-containing, or potentially lead-based particulates from ground-up thermoplastic striping into receiving waters or ephemeral environmentally sensitive areas, these practices are or no longer should be permitted.

I have seen many stockpiles of asphalt grindings stored, unprotected, next to receiving waters that were being used for roadway maintenance and construction. Now, for the most part, at least in California, these practices are not allowed. There are areas that I drive by along interstates that were portland cement concrete (PCC) recycling areas for large highway rehabilitation projects. These areas still today, years after the projects are done, are incapable of growing grass or even weeds, because the soil has been inundated with high-pH lime and PCC by-products. Attempts to hydro seed and amend the soil have failed. Hopefully we can learn from this condition and not allow it to happen again on our watch.

The use of diesel fuel for a bond breaker for paving equipment and hand tools should no longer be implemented, but it is. Paving crews can use environmentally friendlier orange oil and other natural bond breakers so that hydrocarbon-infused chemical products are not entering the soil or being washed out of asphalt during storm water rain events.

Drains, drainages, and drain inlets should be protected or covered during paving operations. While I was performing independent assurance activity inspections on a project, the paving crew did not cover up the drain inlets properly, allowing tack oil emulsion and hot asphalt cement into the drain inlets. These practices need to stop, and environmental stewardship should be recognized as the facilitator for our future generations to enjoy the water quality and wildlife that we are responsible for maintaining into perpetuity whenever possible.

Recently on a large construction project, during a PCC bridge pour over a live creek, the form falsework gave way, causing the forms, construction workers, and fresh concrete to fall into the creek and creek riparian zone. These things happen, but they are preventable. We need to prevent the pollution (and danger to human and all living things) that we are responsible for and can do something about.

During paving and grinding operations, protect the area outside the roadway footprint whenever possible. Don't allow lead-based thermoplastic striping grindings to be comingled with other materials or left on the shoulder of the roadway, but mitigate them properly per WM-4 Spill Prevention and Control and WM-6 Hazardous Waste Management best management practices.

When saw cutting is performed, a wet slurry vacuum must be on site at all times to vacuum up the slurry created from the saw cutting operation. The fine particulates created by the saw cutting process can contain chemicals and elements that will deleteriously affect water and storm water quality if the saw cutting effluent is not mitigated. Additionally, the saw cutting contractor should have a proper disposal site for the collected cementious slurry effluent.

When paving equipment is inactive, it should always be staged over an impervious barrier so that hydrocarbon-based materials cannot enter the soil. This means that during storm water rain events, paving machines should be tarped as well as parked on an impermeable surface. I have seen paving machines parked over plastic that fills up with rainwater. After the storm, a full batch of hydrocarbon-laced effluent enters the soil. This is an unacceptable, egregious practice. Extrapolate the results of BMP implementation in the field. Due diligence in this area will go a long way in implementing effective compliant BMP practices, avoiding potentially costly noncompliance mitigation issues that arise out of deficient BMP implementation.

NS-4 Temporary Stream Crossing—Depending on the applicable permits for the project, a stream crossing may be a very simple BMP implementation, or it may be very involved. If there will be work required in the active flow or riparian channel to install the stream crossing, most likely there will be a 401 Water Quality Certification, 404 US Army Corps of Engineers, or a 1600 series Department of Fish and Wildlife Certification permit. If no actual work is performed in the active flow or riparian channel, these permits may not apply. Make sure that the activities performed are covered under any applicable permits. If water must be diverted for the stream crossing, additional NS-5 Clear Water Diversion BMP implementation may also be required.

Once the temporary stream crossing (NS-4) is installed, be sure that there is no chance for solid waste, silt, sediment, construction waste, debris, or chemicals to enter the water body at the stream crossing point. The crossing should have/be an impermeable barrier between the traveled way and the receiving water. There should be impermeable berms or curbs along the stream crossing so that construction products, wastes, or non-storm water discharges can enter the receiving water. And remember, a dry creek bed is still a receiving water by definition and must be protected as such.

Temporary stream crossing design should be of preeminent importance. Know the area and the historic flows of the riparian or littoral area being crossed. The stream crossing should also be constructed of permit-compliant materials that can be easily cleaned up and removed when the stream crossing is no longer required.

Be aware if there are any environmentally sensitive or archaeological areas or plant species that may not be disturbed for the NS-4 installation, even if these items are not called out on the plans. Also, many of the 401 and 404 permits do not allow fueling of equipment within 50 to 300 feet of a riparian zone. Be sure to know that the project plans and drawings are in compliance with the permits. Review submittals of NS-4 materials used and make sure that they are in compliance with the applicable permits and scope of work. Manage your risk, contractors, and owners alike!

NS-5 Clear Water Diversion—A clear water diversion (NS-5) is a means or method of intercepting clear surface water (non-storm water) on a project and conveying it around or away from the active work area without degrading the water quality of the rerouted water. It may be used in an active water body to isolate construction or soil disturbing activities. Materials implemented in NS-5s include but are not limited to pipes, flumes, sheet piles, gravel bags or berms, turbidity curtains, aqua bladders or barriers, drainage swales, diversion ditches, and k-rail.

The same permit and material compliance is required for the NS-5 as the previously examined NS-4, so apply NS-4 means, methods, materials, and applicable permit compliance components to NS-5 BMPs also.

If pumps are implemented, follow the NS-2 Dewatering guidelines congruently with the NS-5 guidelines as applicable. If pumps are used in the riparian channel or zone that use hydrocarbon-based products, the pumps must be on or over impermeably bermed impermeable barriers or surfaces so that hydrocarbon-based materials cannot enter the soil of the NS-5 BMP area.

A clear understanding of acceptable means, methods, and materials is a must when implementing an NS-5 BMP. Know what is allowed and acceptable, and more importantly, know what is not acceptable!

NS-6 Illegal Connection and Illegal Discharge Detection Reporting— Oftentimes NS-6 BMPs refer to illegal discharges or connections on construction sites by sources other than the contractor. However, as previously mentioned, I was inspecting a site in Northern California where the contractor was discharging dewatering effluent into a municipal publically owned treatment works (POTW) sanitary sewer without a metered discharge point. This was an egregious act by the contractor, and he should have been fined by the municipality. The fact that the contractor was pumping effluent discharge into the POTW was not the offense, but the fact that he was doing it unmetered and undocumented.

Many construction project sites use water from fire hydrants, or indigenous water sources such as lakes, streams, or water tanks. The use of these resources may be part of the permit compliance procedures, and if they are not, caveats apply. There are times when contractors pay private residents or private entities to use their water or electricity for construction activities or their lands for stockpiling construction materials. This is fine, but if you are the owner of the project, you want to make sure that there is a "hold harmless" agreement between the contractor and the private party that you, as the owner of the project, are indemnified, not liable, and separate from the contractor and his activities.

If these types of activities and relationships are not transparent, it can turn into a litigious nightmare. For instance, if a contractor is stockpiling excavated soil from a project site on someone's private land and it is later determined that the soil contains SOCs, VOCs, or aerially deposited lead, the private land could become a Superfund mitigation site and/or a SWPPP nightmare. Contractual due diligence, transparency, and risk management in these cases is the key. Don't create an illegal connection or discharge.

How many times have you left a project site on Friday, only to return on Monday to find that someone has dumped his or her trash or old appliances on the site? This is an illegal discharge, and oftentimes the owner of the project pays for this type of mitigation, depending on the contractual language and the project special provisions if applicable. On the other side of the coin, if a water truck driver discharges a partial load of construction water into a live storm drain, this, too, is an illegal, non-storm water discharge and should be documented and reported. Why? Because it is an unapproved, non-storm water discharge. Municipal water systems are treated and chlorinated for public use. If treated water is discharged into a viable receiving water ecosystem, the treated water could kill aquatic life or change the naturally occurring composition or pH of the receiving water.

Document illicit and illegal dumping and discharges when they happen. Photos can be worth a thousand words! Take photos with the date and time on them, if possible, to corroborate the documentation of the incident, cleanup, and/or removal and mitigation.

NS-7 Potable Water/Irrigation—Remember that we're dealing with *temporary* BMPs here, not permanent ones. Temporary irrigation on a project site may be for dust control, soil stabilization, or temporary seed/plant establishment. The key to NS-7 implementation for irrigation is to ensure that there is no runoff that leaves the site or conveys silt or sediment into active drainages or receiving waters. The water source used for this activity should be free of toxins, sediment, or chemicals that may be deleterious to the foliage or living organisms it comes in contact with.

Many of the receiving waters of California are tainted or impaired with total maximum daily load (TMDL) levels of diazinon, a pesticide, due greatly in part to irrigation of established foliage where the diazinon was applied. Make sure that the portable irrigation equipment that is being used has not been used previously to mix and spray chemical toxins. Portable spray/irrigation units used for chemical applications should not be used for potable water/irrigation tasks. Any water, portable tank, or truck used for non-potable applications *must* be labeled "non-potable" by law.

With irrigation and/or potable water lines, there may be times when discharges are required and allowed. Be sure to know the difference and know what is in the discharged water and where it is going during and after the scheduled discharge. When potable water or irrigation lines are tied into existing services or being maintained, flushing and draining these pipes is common. However, when potable water main work is being performed on existing lines, after the work is complete, the lines are "sterilized" with higher-than-normal concentration amounts of chlorine to kill any bacteria incursion during the construction process. Do not let this chlorine-rich sanitation antibacterial flush water enter into existing receiving waters or drainages. Plan for this and do it right.

NS-8 Vehicle and Equipment Cleaning—For most projects, construction vehicles and equipment are not cleaned on the project site. But if they are, the SWPPP should have very specific language as to how this BMP shall be implemented to ensure NPDES and other applicable permit storm water quality compliance.

According to most project special provisions and permits, all effluent from the cleaning process must be collected and properly mitigated. This can be challenging on remote or small project sites. Most, if not all, effluent discharge from the cleaning process must treated as hazardous or toxic material waste due to the likelihood of hydrocarbon-based fuel and lubricant usage. The presence of silt or sediment are to be reckoned with and must be treated the same as the hazardous effluent if comingled. The effluent must be 100 percent contained and cannot enter the soil or leave the equipment washing area.

There are products on the market that contain, digest, and recycle 100 percent of the effluent fluids. I visited one such manufacturer in Salt Lake City, Hydro Engineering, located on the World Wide Web at http://www.hydroblaster.com/. They offer zero effluent solutions to equipment cleaning for large or compact project sites, including permanent municipal sites. I condone this type of vehicle and equipment cleaning for both temporary and permanent applications for environmental stewardship. If there is any hazardous effluent waste by-product, it is minimal and may be disposed of properly via numerous manifested, chain of custody, mitigation companies that make regular or scheduled pickups of such hazardous materials. Document the cycles of means and methods from beginning to end. Follow SWPPP NS-8 substantive language, and amend it if necessary for compliance.

Quite often in SWPPP language, daily documented inspections must take place when designated activities are performed. Vehicle and equipment cleaning is one of these activities. Technically, when any NS-8 activities occur, there should be a documented daily inspection of the activities performed and the status of all applicable BMPs.

NS-9 Vehicle and Equipment Fueling—Depending on the applicable permits for your project, fueling must be performed from 50 to 300 feet from drains, drainages, or water bodies, unless special circumstances and BMPs are implemented. Know your permit and special provision language. During the fueling process, there is supposed to be someone present at all times at the nozzle while fuel is being pumped. So many times the contractor's fuel truck operator sticks the fuel nozzle into the fuel tank and performs other fluid checks or tasks while the fuel tank is filling. Many times I have seen the result of this practice, seeing the spilled fuel in the soil or water around the fueling area. This is unacceptable. Automatic fuel nozzles do not always shut off.

All fuel nozzles are supposed to have an auto shutoff feature so that when the tank fills to the top, it automatically shuts off the nozzle and the fueling process. But so often, the construction worker puts a rock, tool, or piece of wood in the nozzle handle so it will pump unattended, but then it can't shut off when the tank is full. These practices are out of compliance with environmental stewardship and ethical work practices.

All fueling is supposed to take place in designated fueling areas with spill control and containment BMPs present during fueling operations. I remember reading an older spill control BMP manual. It didn't spell out *preventative* practices for spill control; it spelled out how to clean up a spill. Let's teach spill *prevention control*, and hopefully we won't need the spill cleanup and mitigation. Fuel tanks are to be double walled or located in watertight secondary containment to prevent leaks, drips, or spills from entering the soil or surface of the fueling area.

Observing fueling during asphalt paving operations shows you where the divots in the AC paving will be in a short while. When the diesel or gasoline enters the new asphalt, it degrades the asphalt cement, causing the aggregate not to bond monolithically. A short while after traffic is on the newly paved section, the aggregate unravels, and there are holes or divots in the surface where fuel was spilled during paving operations. Control your spills.

Fueling over water on a trestle platform or bridge takes extra precautions. All fueling should be performed over an impermeable barrier with impermeable containment berms or curbs so that leaks or spills cannot enter receiving waters. I have seen projects where I repeatedly recommended impermeable barriers on trestle platforms over receiving waters, and the recommendations fell on deaf ears due to the cost of the impermeable barrier implementation. It wasn't long after that that the contractor was using hydrocarbon booms and absorbent pads to mop up diesel fuel from the receiving water that the diesel fuel was spilled in. The fueling hose had a splice in it to reach a crane's fuel tank. The splice came apart, dumping diesel fuel onto the trestle platform, where it proceeded to fall into the water, because there was no impermeable barrier on the work over water trestle platform deck. The contractor's risk management plan didn't work well for them that day. You will either pay for compliance or mitigation and compliance both. Manage your risk cost effectively.

Quite often in SWPPP language, daily documented inspections must take place when designated activities are performed. Fueling is one of these activities. Technically, when any fueling occurs, there should be a documented daily inspection of the activities performed and the status of all applicable BMPs.

NS-10 Vehicle and Equipment Maintenance—NS-10 parameters are very contiguous to NS-9 fueling BMP parameters. Maintenance activities should be performed where fluids and by-products of the maintenance activity performed do not enter the soil, drains, or drainages or get left on the surface of the maintenance area. The most common NS-10 deficiency I see in the field is globs of grease in the dirt. Usually at the end of the shift, construction equipment is parked in a line, and the fueling and maintenance personnel fuel and grease the equipment in place, which is out of compliance with most SWPPP NS-9 and NS-10 substantive language. Most SWPPPs state that there shall be designated areas for fueling and maintenance. Make the SWPPP language clear as to the locations and BMPs that shall be designated for NS-10 implementation. Amend the SWPPP if necessary of these BMPs or locations change.

Drip pans, absorbent pads, and impermeable barriers may be implemented, but too often are not. Maintenance work is supposed to be performed over an impermeable dedicated work area. So, put in your SWPPP NS-10 language how you plan to implement the BMP properly and perform equipment maintenance as specified in the SWPPP language. It is very common to inspect the area where equipment is parked at the end of shift for minor maintenance and fueling, to see areas where hydrocarbon-based fluids or solid waste have entered or been deposited on the soil.

Leaking equipment should be repaired immediately or removed from the site. On US Army Corps of Engineers projects, all equipment must pass a safety and proficiency checklist by the QA/QC personnel on-site before it is allowed on the project. Make sure that equipment is sound before sending it out into the field if possible. If a hydraulic hose breaks, it doesn't take long for an excavator, dozer, or loader to pump out twenty or thirty gallons of hydraulic fluid onto the surface of the work area, and that's just the beginning of the problem. The cost to mitigate the fluid spill will most likely exceed the cost of replacing and installing the broken hydraulic hose or fitting.

There should always be a labeled hazmat drum on-site for emergency spill clean-up waste products and for the disposal of hydrocarbon-based waste materials. Do not mix hazardous waste materials with solid waste. This will be referenced later in the WM-6 Hazardous Waste Management BMP language. Batteries, new or old, should be stored in watertight acid-resistant containment. All other hazardous waste materials should be stored in watertight, labeled containers, secured from construction traffic or public access.

Quite often in SWPPP language, daily documented inspections must take place when designated activities are performed. Vehicle and equipment maintenance is one of these activities. Technically, when any NS-10 activities occur, there should be a documented daily inspection of the activities performed and the status of all applicable BMPs.

NS-11 Pile Driving—In the construction industry, pile driving operations consist of but are not limited to driving sheet piles, "H" beam piles, "I" beam piles, Cast-in-Steel-Shell (CISS) piles, steel pipe, cast-in-drilled-hole (CIDH), and concrete piles, to name a few. Most of these operations are performed by similar means and methods, using cranes or excavators and vibrating or pneumatic percussion heads for driving the piles, or rotary drills for drilling CIDH pile holes. The nature of pile driving in construction is usually required because of unstable ground or hydraulic subterranean conditions, which often means working in wet, environmentally sensitive riparian areas or water bodies.

The pile driving BMPs implemented will be determined by the means and methods required by the scope of work and the permits and location of the project. Regardless, NS-9, NS-10, and WM-4 Spill Prevention and Control BMPs will be required at a minimum. If dewatering is required, NS-2 parameters will be instituted. If there is solid or hazardous waste generated, as there often is, then WM-5 Solid Waste

Management and WM-6 Hazardous Waste Management BMPs will be required also. If drilling polymers are required for CIDH or CISS hole drilling and stabilization, then WM-10 Liquid Waste Management may need implementation also.

The key is to know what type of pile driving will be required for your project. Know the bore logs, pile submittal requirements, and geotechnical data so that means and method implementation can be anticipated. Put all the information in the SWPPP to cover all of the potential methods of pile driving BMPs that will be required. It's better to include more potential BMPs than you will need. Just because the BMPs are in the SWPPP, they don't have to be used. The SWPPP can always be amended by the QSD and the owner, but sometimes there can be several days' turnaround time before the amendments are signed by all parties. Often BMPs may be implemented before the amendment is executed with a signed letter of direction or approval by the resident engineer for interim documentation of forthcoming implementation.

Pile driving can be an environmental hot-button issue in some circumstances. Know the work restriction windows and permit allowances when working in or around littoral or riparian water bodies, *and put them on the work window restriction schedule.* Often piles may not be driven in water bodies during times of aquatic life spawning or migratory activities. When working in or over receiving waters, sometimes vegetable oil use is required over regular hydraulic oil for pile hydraulic driving equipment. Have all spill control BMPs available and an up-to-date spill prevention and control plan. Have all the required notification numbers to be notified in the event of a spill. This is required. Get the real-time numbers; don't use outdated templates.

Contain and control everything that can be contained and controlled! I have seen CIDH and CISS pile operations in riverbeds where the drill rig was flinging grease over the entire work area while performing pile driving activities, which happened to be in the actual dry riverbed. There was grease spatter all over the native river run cobble, sand, and gravel.

Pile driving can be an oily, greasy, smelly operation. Place an impermeable barrier over areas that require protection, or be prepared to mitigate the affected area in a timely and expensive manner.

Spend a little extra time conferring with your construction personnel and, if necessary, the owner of the project, and try to get this BMP right the first time. There is a lot of risk at stake here potentially. If an impermeable barrier for the work surface should be used over water, buy it and install it correctly. Know your means, methods, and risks, and manage them thoughtfully; they're like fresh concrete—you work it smartly, or it will work you and you may not get the desired end.

Quite often in SWPPP language, daily documented inspections must take place when designated activities are performed. Pile driving is one of these activities. Technically, when any pile driving occurs, there should be a documented daily inspection of the activities performed and the status of all applicable BMPs.

NS-12 Concrete Curing—Concrete curing environmental concerns will be material storage, spill control, and application rates and methods. Most of the NS-12 deficiencies that I've seen are spill control and material storage. The actual curing compound should always be stored in watertight secondary containment that is protected from damage by construction operations and traffic. On large projects the cure compound often comes in 500-gallon high-density polyethylene (HDPE) containers on a pallet with a shutoff valve at the bottom of the container. While my point of view here has been unpopular, I feel that these containers should be stored in some type of secondary containment. If the HDPE containers get punctured, degrade due to ultraviolet contact, or the valve leaks for one reason or another, this material should be contained so that it doesn't enter drainages or the soil. If the cure comes in 55-gallon drums, store them in watertight secondary containment so that storm water cannot build up on the top of the drums.

When applying concrete cure, it should be applied per specification, not too thin nor too heavy. It should be protected from rainwater and wind events so that it cannot enter nearby soil or receiving waters. If a concrete pour finishes just before a rain event, cure should not be applied; a cover-in-place method should be implemented. Plan ahead. Storm water can convey uncured materials (cement, concrete, and curing compound) to soil or receiving waters, initiating a spill/cleanup hazmat situation.

Water curing can be problematic if not planned and implemented properly. It is a fine balance to implement water curing per plan specifications and still have zero effluent. If effluent by-product results from curing operations, have a plan so that it won't affect water quality or storm water management plan specifications. Due to the porosity of concrete, effluent from concrete curing can convey high-pH effluent containing other PCC chemical residues and admixture chemicals. Keep water cure systemic application effluent controlled and contained according to NS-12 BMP storm water management plan specifications. Manage risk.

NS-13 Material and Equipment Use Over Water—Due to the unforgiving nature of a mistake while performing this type of work, I suggest taking it very seriously—not only from a BMP risk management point of view, but from an environmental stewardship point of view equally. Just the fact that work is being performed over water implies that there are most likely 401 Water Quality Certification, 402 NPDES,

404 US Army COE, and 1600 series Department of Fish and Wildlife permits involved, and possibly more.

The checklist for writing the SWPPP plan for this BMP must be all-encompassing, employing all necessary and probable BMPs, as well as *entering environmental and wildlife work window restrictions in the project schedule*. Material, task means, methods, scope of work, and procedure submittals will dictate potential BMP needs and implementation to avoid fatal flaws. This information must all be weighed and factored in to a complete, professional SWPPP. Corresponding BMPs that will most likely accompany NS-13 may be but are not limited to most of the non-storm water management (NS) and waste management (WM) SWPPP BMPs. Since work over water locations are often small, much thought must go into how to best orchestrate and implement all the required BMPs and still have room to work!

The first thing that comes to mind about material and equipment work performed over receiving waters is, however possible, to implement an impermeable barrier between the work surface and the receiving water below. Project special provisions and standard specifications often refer to watertight curbs or toeboards, but they never come right out and specify the use of an impermeable barrier. What good is an impermeable curb or toeboard without an impermeable barrier under it? The impermeable barrier is *implied, not specified*. Determine the risk, and if there's a chance for being fined for a spill into receiving waters, spend the money for the impermeable barrier.

This can be challenging because the impermeable barrier must be protected from construction activities and must remain intact to be effective. I have seen 10 mil and thicker membranes placed under plywood or timbers so that the membrane is not damaged. Or the membrane could be placed under the flooring or falsework of the deck. If there are fragile aquatic systems or receiving waters below the work deck, the cost of the impermeable membrane is money well spent. I prefer the membrane to be installed as close to the top of the work deck as possible so that the requisite watertight curbs or toeboards can be installed in conjunction with the impermeable membrane.

One challenge regarding the impermeable membrane is storm water buildup during rain events. If the project is in an area of high rainfall or runoff, a potential overflow discharge could occur from the impermeable deck work area. If this is an issue, cover as much of the deck as possible during rain events and install an overflow catchment to capture the overflow water. If there have been construction chemical or material spills on the deck area, the overflow water will most likely have to be conveyed to a holding tank, where it can be characterized for chemicals, pH, and turbidity prior to discharge. This would be an extreme worst-case scenario situation. If possible, an alternative is to filter the overflow water through a series of absorbent pads or hydrocarbon booms and let the water discharge where it cannot cause pH, turbidity, or erosion issues from effluent discharge.

Another challenge of working over water is mitigating solid waste. Protect all solid waste receptacles with lids and watertight/windproof protection. As soon as solid waste is placed into a receptacle, the lid should be replaced unless it is not windy and solid waste mitigation activities are in progress. All solid waste containers are to be watertight at the end of every work shift or at the end of the day, as applicable.

Quite often in SWPPP language, daily documented inspections must take place when designated activities are performed. Work over water is one of these activities. Technically, when any work over water occurs, there should be a documented daily inspection of the activities performed and the status of all applicable BMPs.

NS-14 Concrete Finishing—This BMP is for both freshly poured concrete and for existing concrete that is being refinished or being finished for the first time after being fully cured. The obvious applies here. Do not create any effluent from high-pressure water or air that isn't properly handled or disposed of.

Fresh concrete will most likely contain lime, hexavalent chromium, calcium silicates, and alkalis that can be hazardous to unprotected skin, not to mention aquatic life. Salts from organic acids from air-entraining agents can create alkalinity issues for pH readings. Airborne dust from set-up concrete will contain remnants of admixture chemicals and crystalline particulates of silica and, most likely, more airborne deleterious agents.

Whether finishing fresh or set-up concrete, the BMPs are about the same: don't discharge by-products from the work area, contain effluents or waste, and mitigate waste or effluent properly.

Remember, PCC waste should not come in contact with native waters or soil. The alkaline qualities of the waste or effluent are toxic to human, aquatic, and wildlife. Give a hoot; don't pollute!

Quite often in SWPPP language, daily documented inspections must take place when designated activities are performed. Concrete work and finishing is one of these activities. Technically, when any concrete work occurs, there should be a documented daily inspection of the activities performed and the status of all applicable BMPs.

NS-15 Structure Demolition/Removal Over or Adjacent to Water—The same parameters apply here as to NS-13 BMPs. This work is usually heavily permitted.

Demolition debris must be kept from receiving waters and from leaving the site via storm water or airborne conveyance. When paint is stripped from structures, it must be contained and vacuumed up and will most likely be mitigated as hazardous waste. There may be environmental or fish and wildlife work window restrictions as well, so be sure to get any restricted time periods on the project schedule for critical path management.

CHAPTER 11

Waste Management Best Management Practices (10)

Waste management consists of implementing procedural and structural BMPs for handling, storing, and disposing of construction materials and wastes generated by the project to prevent the release from the site of construction or waste materials via airborne or storm water discharges.

WM-1 Material Delivery and Storage—All material delivery and storage (WM-1) areas must be noted in the water pollution control drawings. Update the water pollution control drawings as often as new changes are made or removed and at least monthly by default. The WM-1s must have perimeter sediment control BMPs (usually fiber rolls or silt fence) and, if necessary, environmentally sensitive area fencing to delineate inactive project areas from project active construction areas. The storage site must also be protected from *run-on* water, if applicable, from any area up gradient of the site. The WM-1 area is usually considered an active or disturbed soil area unless entirely on asphalt, concrete, or other impervious surface. Many WM-1s have TC-2 or TC-1 stabilized construction roadway or temporary construction entry/exit BMPs installed. There should be no material or sediment tracking from the WM-1 area.

In most project special provisions, when hazardous materials are delivered and consequently stored at a project delivery site, the material delivery itself must be inspected, and the site storage area should be inspected daily while hazardous materials are present. The person receiving the hazardous materials should be trained in hazardous material handling and spill prevention and control.

Many WM-1 storage areas are a target-rich environment for deficiencies due to the multifaceted nature of the materials stored there. There is often hazardous waste, solid waste, fuel and fluids for equipment, and hazardous construction fluids and chemicals present, as well as construction vehicles and equipment. Oftentimes there are stockpiles at the WM-1 areas. All stockpile areas are also to be noted on the water pollution control drawings, even if the stockpiles are within the boundary of the WM-1. Stockpiles must be protected per WM-3 Stockpile Management storm water management plan (SWMP) specifications, which will be reviewed shortly under WM-3.

Hazardous waste and non-waste fluids must be stored in watertight secondary containment. A conex storage container with a wooden floor is not a secondary containment apparatus. If a five-gallon bucket of oil or paint spills near the door of the storage trailer, the fluid will escape into the soil or whatever surface the storage container is on, or the spilled fluid will be tracked by foot traffic outside the storage area. In short, the fluids are not contained. Many times I have seen the flooring of these storage containers rotten or broken, allowing construction fluids to leak out of the container into the soil. Oftentimes the roofs leak and allow storm water to enter the storage area. This can allow saturation of the wood flooring, leading to eventual failure of the structural integrity of the floor, and create opportunity for fugitive materials to enter the soil under the storage area. I have seen oil saturation of wooden floors to where the oil actually penetrates the wood and drips into the soil under the storage area.

All working stock fluids and waste fluids must be stored in watertight secondary containment, so if they escape their packaging, they will be contained from exiting the storage area. Batteries or acidic chemicals must be stored in watertight acid-resistant storage apparatus. This is a health and safety, as well as an environmental, issue. All materials should be clearly labeled and stored or protected per Material Safety Data Sheet (MSDS) specifications. Liquids, petroleum products, and substances listed in 40 CFR Parts 110, 117, or 302 shall be stored in approved containers as specified by the applicable DOT. Do not store incompatible materials, such as chlorine and ammonia, in close proximity to each other.

Uncovered secondary containment facilities should have the capacity to contain storm water from a twenty-four-hour, twenty-five-year storm, plus 10 percent of the aggregate volume of all the containers or the entire volume of the largest container in the facility, whichever is greater. Any facilities that capture storm water should be drained, and the effluent characterized and properly mitigated. This effluent water should be considered hazardous unless characterization by an approved lab proves differently.

Stockpiles, pile driving equipment, vehicle and equipment fueling, and washing and maintenance must be performed or located at least 100 feet from concentrated flows of storm water or receiving waters and drainages if within a floodplain, and at least 50 feet if not in a floodplain, unless otherwise authorized. This goes for portable toilets, too. Know about the storage area physical features per US Government Survey (USGS) as related to floodplain details, if necessary.

Stored materials such as cement or concrete should be protected from wind, water, and construction traffic. All solid waste containers should be protected from wind and storm water events to prevent fugitive solid waste. Often metal debris or solid waste Dumpsters have holes in them or leak out the doors. Storm water incursion can cause fugitive sludge discharge from the Dumpsters that may deleteriously affect water quality if discharged near drainages or receiving waters.

Treated lumber, railroad ties, or similar items cannot be stored directly on the ground and must be covered or protected by an impermeable membrane and protected from storm water contact. A labeled hazardous waste 55-gallon drum should be on-site at the storage or project area for incidental hazardous waste cleanup disposal and be manifested with a chain of custody to the applicable certified disposal facility.

WM-2 Material Use—The material use (WM-2) BMP is essentially about identifying the materials, fluids, chemicals, and such that are being used daily on the site for the construction process. Know the Material Safety Data Sheet (MSDS) properties of the product being used and minimize the potential of these products to enter the air, soil, MS4 drainages, or receiving waters. It is important that these materials be labeled or properly identified for code of safe practices and water quality issues.

It is important when certain materials—such as materials containing acids or other volatile chemicals—are used that they are only implemented by appropriately trained employees. When spraying paints or solvents, protect the applied materials from wind and storm water events so they do not become fugitive from the work area and enter the water or soil or leave the site.

I was on a paving project where the tack oil was sprayed on the highway prior to a final overlay lift. While traffic was being cleared in the other lane, a cloudburst hit the tack oil, sending it into the lane of live traffic, coating the vehicles with tack oil, and some of the storm water effluent entered existing drainages. Manage the risk. There were many costly claims over this incident.

Be sure to have adequate containment and spill cleanup and control materials near the material use if applicable. Preventing a spill is always better than cleaning up a spill. Be proactive and avoid having to be reactive whenever possible. Place impermeable covers, barriers, or check dams prior to material use. Pile driving can send oil and grease spatter for a hundred feet or more. When spraying paint or similar materials, protect drainages, drain inlets, and personal and private property. I have seen temporary roadway shoulder striping that goes right up to a drain inlet metal grate and right over it. This is unacceptable. Plan for material use and implement its proper BMP application(s).

WM-3 Stockpile Management—Stockpile management (WM-3) is definitely one of the most deficiently implemented BMPs on the sites that I have inspected. The basic rules for WM-3 implementation are pretty standard for the construction industry. A stockpile is either active, inactive, hazardous, or nonhazardous.

A stockpile is considered inactive if it has not been disturbed for more than fourteen days. Inactive stockpiles are to be protected per applicable WM-3 specifications after fourteen days of inactivity; it doesn't matter if there are no forecasted high winds or rain events, the stockpile must be protected.

A "protected" nonhazardous material stockpile means that there are perimeter controls completely around the base of the stockpile as a linear sediment barrier (usually silt fence or fiber rolls). The stockpile material must be covered with a permeable or impermeable barrier, or stabilized with hydraulic mulch or binder fiber matrix (BFM), to protect against wind, raindrop, or storm water erosion.

All nonhazardous material stockpiles must be "protected" within seventy-two hours of a predicted storm water rain event or high wind event (depending on the stockpiled material and particulate size, particulates can become airborne when winds exceed 13 mph). Depending on where your project is located, a forecasted "qualifying rain event" is a forecast of a 50 percent chance or more of 1/10" of rain or more. Know the standards for what constitutes a forecasted "qualifying rain event" where the project is located. This will be located in the project special provisions or federal, state, or regional permit. Follow the most stringent definition for assured compliance.

There are specific means and methods for the implementation and installation of WM-3 BMPs. Seldom are these BMPs installed correctly. Silt fence and fiber rolls must be installed as described in Chapter 7 per the SC-1 and SC-5 specified data. Temporary cover must follow the WM-3 specifications.

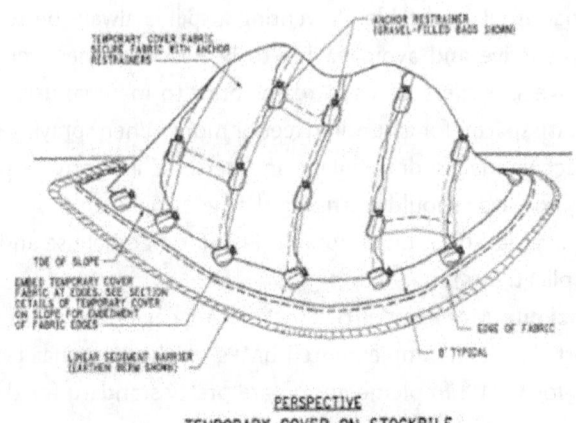

PERSPECTIVE
TEMPORARY COVER ON STOCKPILE

Image from *California Department of Transportation 2010 Standard Plans*, page 248.

The perimeter control must be away from the cover, not on top of it. Oftentimes the fiber rolls are holding down the plastic. This is an unacceptable practice. Per the drawing above, the cover should be properly and adequately wrapped and lapped to create a shingle effect for storm water to run off the pile. The lapped fabric should take into account prevailing wind directional flows, if applicable. The rock bags and rope should be made of DOT-specified compliant materials or equal regarding ultraviolet and impact/tear strength-resistant material specifications. Very seldom have I seen these BMPs installed or maintained per storm water management plan (SWMP) specifications.

Hazardous Material Stockpiles

There are at least four types of hazardous material stockpile characterizations.

1. Treated wood products are to be placed on a pallet or dunnage and stored completely protected from storm water with no earthen contact.
2. Asphalt cement cold mix should always be stored on an impermeable surface where the hydrocarbons from the cold mix cannot enter the soil or come in contact with storm water. AC cold mix must be covered with an impermeable cover before rain events or after fourteen days of inactivity.

3. Hazardous soil, if containing aerially deposited lead (ADL), should be covered at all times when inactive, especially during wind and rain events. The ADL soil should never be mixed with other non-ADL soil unless, per the project plans and specifications, sometimes ADL soil with low levels of lead may be buried in subgrade lenses where the soil will be covered by an impermeable surface or treatment.

4. Soil containing hazardous chemicals that could leach into existing soil should be placed on an impermeable surface and covered with an impermeable barrier when inactive, and always be covered before forecasted wind or rain events. These piles should be marked so that all parties are aware that the soil is contaminated. After this soil has been characterized at a lab, it will be manifested and hauled to the appropriate disposal site with chain of custody documentation.

Stockpile management is now a full-time BMP implementation commitment. With the new federal and state construction permits, there is no more defined rainy season. It is now important to consider, design, implement, and maintain all types of stockpile management year-round.

WM-4 Spill Prevention and Control—A lot of spill prevention and control (WM-4) BMPs have been addressed in the previously described BMPs, including but not limited to vehicle and equipment washing, fueling and maintenance, work performed over water, pile driving, and material storage and use. These BMPs are integrated into spill prevention and control. The WM-4 BMP is primarily associated with spills originating from the contractor and its subcontractors building the project, not with spills generated from the public or accidents not related to the project. This BMP is project- and contractor specific.

The SWPPP substantive language for the spill prevention and control BMP should be descript and thorough. This is not the time for "copy and paste" language from an old SWPPP plan. All emergency contact phone numbers must be listed in the SWPPP language, including twenty-four-hour contractor personnel contacts. Either the QSD or a designee for the spill prevention and control plan should be hazmat trained and certified so he or she will know how to react as a first responder and escalate emergency resources expeditiously and professionally as necessary. Some project special provisions require either the QSD, water pollution control manager, or a hazmat-trained and certified designee to be available 24/7 during the project duration. If a substantial spill occurs, the reaction and implementation time is critical for damage control and, if necessary, for employee and public safety. The project resident engineer should be notified immediately, or as soon as is practical if the spill occurs outside of regular working hours.

There are basically three types of spills: minor, semi-significant, and significant or hazardous.

A *minor* spill consists of conventional working stock or non-highly toxic materials that are small enough to be controlled by the contractor's first responder. If the spill can be contained, the material recovered, the area cleaned up without designated hazmat personnel, and the materials properly documented and disposed of, it is a minor spill.

A *semi-significant* spill is the same as a minor spill except it may take several contractor personnel under the direction of the contractor hazmat designee to mitigate and properly dispose of the spilled and affected materials.

A *significant or hazardous spill* cannot and should not be mitigated or controlled by jobsite personnel. Only trained and qualified personnel should deal with significant or hazardous spills. Immediate notification of the appropriate emergency services is required, including local, state, and federal officials as required per state emergency management agency warning centers and the National Response Center criteria, if estimated measurable quantities of spilled material meet the quantities cited under 40 CFR 110, 119, and 302. Local, state, and federal responders may vary per the contractor's hazmat designee's and, if applicable, the area 911 dispatcher's discretion.

The best spill control is obviously spill prevention. However, have the appropriate materials for cleanup and mitigation on-site in the proper amounts proportionately to the materials potentially requiring mitigation. Train employees about spill prevention and control weekly and how to deal with manageable spill situations by not spreading or conveying spilled materials and the implementation of means and methods of safely containing the spill and affected area in minor and semi-significant spill situations.

WM-5 Solid Waste Management (also known as Solid Waste Disposal and Recycling)—The main things to remember regarding WM-5 BMPs is that the solid waste receptacles (bins, Dumpsters, boxes made on-site, trash cans, etc.) must be protected from wind and storm water incursion. What good are Dumpsters with lids if they are left open during wind or rain events? Good housekeeping practices are important, or the solid waste will never make it to the Dumpster! And once the solid waste makes it to the Dumpster, you want it to stay there until it's removed from the site! All solid waste receptacles should be closed or covered at the end of every shift or during windy or rainy conditions. Almost every project site that I've inspected has had at least one or two unprotected solid waste cans or Dumpsters. This is one of the most common deficiencies. Dumpsters should be located at least 50' from drainages and drains.

Another problem with Dumpsters on-site is that some are for steel recycling, some are for solid waste, and some are for other recycling activities. A tour around the site visiting the various Dumpsters sees that there is steel in the solid waste bin and solid waste or tree stumps in the steel bin. Solid waste and recycling is not that hard to manage, but it takes training and effort by the on-site employees. Someone has to care and be responsible to change the culture of solid waste mitigation. Empty the Dumpsters when they are full, and only order watertight receptacles.

The most egregious environmental problem with solid waste mitigation is that often hazardous waste is comingled with solid waste. This is a problem on most sites. All project special provisions that I have read in the last few years clearly spell out that hazardous waste must be kept separate from solid waste and mitigated separately and properly. How often do you see 5-gallon paint or concrete cure buckets in the solid waste Dumpster? How often do you see empty gallon or quart motor oil containers in solid waste Dumpsters? It's time for being responsible for environmental stewardship and caring for our environment! And it will only happen by raising the social consciousness and culture of our workers and employees by training and educating them to be environmentally responsible. *One gallon of oil can contaminate 1,000,000 gallons of water! One drop of oil can contaminate 75,000 gallons of oil!* Let's start caring and educating at home and on the project sites. Caring starts and ends with the person in the mirror!

It is easy to tell the culture of a project site within the first few minutes in the field, and the same goes for the SWPPP files also. Compliant solid waste management must be taught over and over to all employees, new and existing. This compliance starts with the project management team and the organic culture of the contractor. Whatever materials come to the site in people's lunch box should leave in their lunch box or be securely deposited in a compliant solid waste receptacle. Sounds simple, but it is rare to find on the hundreds of project sites that I've visited.

Looking at solid waste BMPs from a permitting and environmental compliance view, realize that comingling hazardous waste with solid waste is not only a WM-5deficiency; it can be a permit compliance violation that could end up as a fine or citation if water quality is deleteriously affected. If solid waste becomes fugitive from the project site, it is an unauthorized non-storm water discharge. A non-storm water discharge report should document the incident, with a corresponding corrective action report explaining and detailing mitigation actions performed for compliance.

See waste management BMPs not only as SWPPP compliance components, but also realize that noncompliance with waste management BMPs may result in contract fund withholding, citations from the promulgating regulatory permittors, or monetary fines and/or jail time for willful negligence. Many owners of projects require annual or final project solid waste mitigation compliance reports and documentation, or project funds may be withheld up to $10,000. Manage the risk!

WM-6 Hazardous Waste Management—In California, if a project has known hazardous waste generation activities, the water pollution control (WPC) manager or a qualified designee must have successfully completed training under 22 CA Code of Regulations Section 66265.16. The designee must also be familiar with handling and emergency procedures under 40 CFR § 262.34(d)(5)(iii). Other states may have similar CFRs, especially if federal funding is involved in the project.

The WPC manager or designee's qualifications and credentials must be documented in the body of the SWPPP. The hazardous waste manager (HWM) must oversee, inspect, and enforce applicable regulatory procedures for all hazardous waste material mitigation implementation activities including but not limited to the generation, storage, transportation, manifesting, chain of custody, and proper disposal of all project hazardous waste.

Unanticipated Hazardous Waste

Upon discovery of unanticipated hazardous waste, the designated hazardous waste manager (HWM) must immediately isolate the area until the material is identified, the situation evaluated, and the best course of action determined. Notify the RE or owner of the project as soon as is practical. Once the material is identified as defined in Labor Code § 6501.7 or a hazardous substance as defined in Health & Safety Code § 25316 and § 25317, the HWM, contractor, and owner of the project will formulate a mitigation plan. The hazardous material must be identified and characterized by a certified Department of Health Services Environmental Laboratory Accreditation Program (ELAP) or equal laboratory so that appropriate mitigation can be carried out. The National Response Center, at (800) 424-8802, should be notified of spills of federal reportable quantities in conformance with the requirements in 40 CFR parts 110, 117, and 302.

All hazardous wastes must be stored and transported in watertight metal containers as approved by the US DOT (usually labeled 55-gallon drums with bolted lids). The labels should be undamaged and legible, identifying the contents and whether they are toxic or not and the accumulation start date per the applicable state and federal code of regulations. The containers should be stored at least 50' away from drainages and secured from construction and public vehicular and pedestrian access and traffic, that is, a locked chain-link fence enclosure or lockable shipping container. These materials should always be stored in a watertight, secondary containment facility, preferably where storm water cannot accumulate. The storage area, if subject to storm water influence, must have the capacity to contain the precipitation from a twenty-four-hour-long, twenty-five-year storm and 10 percent of the aggregate volume of all containers or the entire volume of the largest container within the facility, whichever is greater. If storm water accumulates in the storage area, it must be mitigated as hazardous waste also, so it is best to store hazardous materials in containment where storm water cannot accumulate. Solid materials should be stored on pallets if stored outside or subject to storm water incursion; liquids should never be stored outside watertight secondary containment.

Do not allow hazardous waste to come in contact with the ground or soil where the mitigation takes place or to be mixed with other waste products, hazardous or not. Do not overfill the containers. Dispose of hazardous waste within ninety days of the start of generation. Use a hazardous waste manifest and a transporter registered with the Department of Toxic Substance Control (DTSC) and/or in compliance with applicable state trooper or highway patrol parameters to transport hazardous waste to an appropriately permitted hazardous waste management facility. If the hazardous dry or soil material is not in containers but is loaded into trucks, the trucks must have fully closable lids or tarps so that none of the materials can become fugitive by wind or storm water erosion. As mentioned earlier, there must be chain of custody (COC) documentation for the transportation and disposal of the hazardous material as noted on the sequentially numbered manifest paperwork that must accompany each load to the operator of the approved hazardous material dump site. The COC documents the origination and the disposal site of the hazardous waste material. The owner or resident engineer (RE) of the project will obtain the US EPA Generator Identification Number for the manifests. After signing the manifests, the RE will provide them to the contractor performing the hazardous material mitigation.

WM-7 Contaminated Soil Management—Contaminated soil management (WM-7) parameters are very similar to WM-6 Hazardous Waste Management, based on the two classifications of materials: anticipated and unanticipated as far as storage and protection. However, there are some differences in the means and methods of dealing with contaminated soil that differ from dealing with hazardous waste.

Anticipated contaminated soil normally has contract bid/pay items for the mitigation of the contaminated soil. Often there is an approximate quantity and characterization of known contaminated soil. These are anticipated, described, and managed per the project plans, specifications, special provisions, and bid item description language.

Larger quantities of contaminated soil, if stored, should be stored on an impervious surface, completely covered with an impervious membrane that is secured per WM-3 Stockpile Management specifications so that it is not subject to wind and storm water erosion. The hazardous soil should be identified with signage so that the material is not accidentally mixed with other materials or disturbed.

Unlike hazardous waste, aerially deposited lead (ADL) soils are often buried in roadway subgrade sections or similar under impervious layers of imported material to seal off and deter lead migration. Soils with high concentrations of ADL are mitigated to hazardous material sites, where they are disposed of per the aforementioned chain of custody/manifest documentation parameters mentioned in the WM-6 language. Prior to performing any known contaminated soil work, the soil shall be tested by a Environmental Laboratory Accreditation Program (ELAP) lab or equivalent laboratory; then contractor and subcontractor employees must complete a safety training program that meets 29 CFR 1910.120 and 8 CCR 5192 covering the potential hazards as identified.

To minimize on-site storage, if contaminated soil shall be transported from the site for disposal, it shall be performed in accordance with all applicable regulations, including but not limited to Title 22, CCR, Sections 6626.250 to 66265.260 per applicable state and federal manifested chain of custody parameters. These materials must be disposed of at a certified Department of Toxic Substance Control (DTSC) or equal certified site.

The main difference between anticipated and unanticipated hazardous soil mitigation is how they're paid for under the contract substantive language. They are usually protected, stored, transported, and mitigated the same, with the same BMP criteria as the WM-6 BMPs.

WM-8 Concrete Waste Management—Concrete waste management (WM-8) is another BMP that often is overlooked and out of compliance. For the sake of clarity in discussing this BMP, let's call portland cement concrete or white concrete "PCC" and asphalt cement concrete, black concrete "AC." Most projects that I've inspected have WM-8 deficiencies or areas of noncompliance conforming to the SWPPP storm water management plan (SWMP) specifications and language. In days past, when we poured concrete, it would end up all around the site of the pour, and if there was a cubic yard or so left over, it was pumped or discharged somewhere for backfill in an excavation.

This cannot be done now without authorization or special permissions. All concrete must go into the designated area being poured, or placed, properly disposed of, or removed from the site in the delivery truck that brought it.

Fresh concrete and its residual and constituent by-products are very ephemeral in the deleterious environmental effect time frame. After a couple of hours, water, cement, chemicals, aggregate, admixtures, oil, and sand become PCC or AC concrete, and in a few days can develop a compressive strength of over 4,000 PSI. But during the ephemeral phase of the mixing, placing, and setting up of the aggregate compound mixture, a lot can happen. Even subsequently, after the concrete has set up, while it is less apt to contribute deleteriously to its environmental surroundings, it can still contribute to the changing of soil pH or leaching measurable amounts of chemical admixture into the soil or receiving waters. At any AC or PCC batch plant, there are warning signs that these products contain chemicals that may cause cancer or serious illness. This may be a clue that these products should be handled responsibly when introduced into a sensitive environmental area.

Before attending World of Concrete convention classes on concrete air entrainment and admixtures, I had never realized that concrete is really a big porous hard aggregation of elements. Water and other liquids actually enter and migrate in and through concrete that is already cured and set up. That's why we put the asphalt/tar emulsion and impermeable barriers on our basement concrete walls and structures, because of the porous nature of concrete, to stop water from the soil from leaching into our homes and turning our cinder block walls into mush. Understanding this fact, one may be better prepared to understand why keeping freshly placed concrete and its by-products from tainting our receiving waters and soil is so important regarding its potential impact on water quality.

The main BMPs concerning PCC and AC concrete are to keep the freshly mixed products out of drains and drainages during the application process, including AC tack oil emulsion, PCC and AC slurry from saw cutting operations, and any other cementitious containing products like mortar and patching compounds.

Proper mitigation and disposal of cementitious products has been a problem on most projects that I've inspected. Since concrete products have been around for hundreds of years, it seems that it is taking quite a while for the construction industry and the public to get serious about cement product mitigation and disposal. Most inspections that I have made on PCC concrete washout and waste disposal storage areas have more often than not revealed notable deficiencies. Concrete mixer truck drivers often miss the washout area when washing their chutes and leave PCC residue outside the contained washout area. There is more of a problem with PCC concrete residue than AC concrete because once the AC concrete residue sets up, it becomes impervious to raindrop and wind erosion, so pH and chemical reactions affecting the soil are negligent. PCC residue may still affect soil and receiving water pH even after its initial setup. A general rule is that PCC and AC waste should not be in contact with soil or impervious surfaces where residues can be conveyed to mature permanent vegetation, drainages, or receiving waters.

PCC and AC concrete disposal and stockpile areas should be protected from wind and raindrop erosion. When applicable, AC and PCC waste stockpiles should be stored on impermeable surfaces and protected during periods of inactivity and wind and rain storm events. When waste storage areas or apparatuses are more than 75 percent full, they should be maintained and the excess materials disposed of properly, and they should be covered with impermeable barriers during storm water events. And remember, most SWPPP plans and/or special provisions require inspections *each day* that AC or PCC concrete is used on the project. Be sure to document means, methods, and deficiencies in a written report with photos and that the camera that is used for documentation has the date and time when the photos were taken. Date and time of day can be critical when recreating deficiency or corrective action reports after the fact.

WM-9 Sanitary/Septic Waste Management—There are several key issues to consider with sanitary/septic waste management (WM-9) BMP implementation.

1. Portable toilets should be located at least 50' from drains, drainages, or receiving waters.
2. Portable toilets should have a secondary containment apparatus under the unit to catch any chemicals or waste that may be discharged from the floor of the unit during maintenance. When the secondary containment fills up with storm water, have the sanitary maintenance company suck the water into their tanker trucks when they are maintaining the units.
3. Portable toilets should be secured from being upended by wind or acts of vandalism.
4. Most Department of Health and Safety regulations require hand-wash stations at portable toilet locations. There must be a watertight solid waste container for disposal of solid waste at these locations. Put some thought into this, and it will pay off.

While performing inspections on projects, I have witnessed chemical discharges multiple times during sanitary toilet maintenance. Technically, this is an unauthorized non-storm water discharge of hazardous chemicals and should be treated and documented as such. The sanitary toilet service personnel often have many units to service daily and often do not pay attention to detail regarding the containment of the waste and chemicals involved with the sanitary unit's maintenance. If temporary sanitation facilities are located in environmentally sensitive areas, be sure to relay information to the sanitation company that care must be implemented during maintenance activities.

Be knowledgeable about the permits on your project. Some permits do not allow temporary sanitary facilities within 100' of receiving water or environmentally sensitive areas.

WM-10 Liquid Waste Management—Liquid waste management (WM-10) applies to several construction site activities including but not limited to:

1. Polymers, slurries, and fluids used in drilling activities.
2. Active treatment systems (ATS) coagulation effluent chemicals.
3. Portland cement concrete and asphalt concrete pouring, placing, and cutting.
4. Sludge from dredging activities.
5. Liquefaction from sediment basins or holding ponds.

For materials generated on the project from project bid items, these materials must be mitigated per project-specific permit requirements. If nonhazardous, they must be protected from wind and storm water erosion until mitigated. If hazardous, waste must be protected from wind and storm water erosion and from contact by construction or public traffic until mitigated. Hazardous materials must be labeled appropriately and have ample notification that the materials are hazardous. Normal contract bid items contain language stating that full payment for mitigated liquid waste will not be rendered until documentation of proper mitigation and disposal of liquid wastes per project-specific permits and special provisions has been provided by the contractor.

Hazardous wastes discovered on the project site will be mitigated per, but not limited to, Title 22 California Code of Regulations (in California) and Section 40 of the Federal Code of Regulation specifications. These materials must be analyzed and characterized by an Environmental Laboratory Accreditation Program (ELAP) certified lab or equal state Department of Public Health lab.

All liquid waste must be stored at least 50' from drains and drainages and 100' from receiving waters or drainages when in a known floodplain.

Part Three

Balancing Risk Management and
Environmental Stewardship

CHAPTER 12

Risk Management with
Environmental Stewardship

After nearly forty years in the construction industry, I have witnessed much change in the means and methods of the construction process. Not until the early 1990s did storm water pollution prevention become prominent and requisite on the heavy construction scene, at least in California, where I was working.

The old way of doing things has gradually given way to a higher degree of environmental conscience and stewardship. Environmental compliance in the construction industry is no longer a subjective choice but a mandated requirement.

Manage Your Risk

As project manager on a project in rural Arkansas, I learned how important environmental compliance can be. As an independent consultant to a large corporate entity, I was in charge of all site construction operations, including quality assurance/quality control (QAQC) and SWPPP oversight. Eventually the entity that I was consulting to realized that we were going to have to fire the general contractor and replace him. In preparation for the litigation, many things were of subjective opinion between the general contractor and my client. But the one thing that made it easy for the arbitrator to award in my client's favor was the fact that the general contractor was way out of compliance with SWPPP issues. I had carefully documented off-site tracking onto public traveled ways due to no temporary construction entry/exit or stabilized construction roadway (TC-1 and TC-2 BMPs). There were multiple e-mails and photos documenting noncompliance with storm water pollution prevention plan requirements per the

Department of Environmental Quality permits for the project. The general contractor's noncompliance with storm water pollution BMPs was ultimately what tipped the scales in the decision-making part of the litigation, where he had no rebuttal argument. The general contractor employed zero SWPPP risk management into his project strategy, and it ended up costing him millions of dollars in profit.

Another issue that came up was that there needed to be several miles of buried pipe installed for the project. I was directed to dig across a nearby river with heavy equipment in the summer when flows were low to install the pipe. I informed the client that it would be taken care of implementing proper means and methods.

I hired a contractor to bore and jack steel casing under the river channel so that no equipment need enter the littoral or riparian zone to complete the requisite piping. Before the project was started, the US Army Corps of Engineers and Department of Fish and Wildlife were aware of the permit application. SWPPP BMP implementation for such a task would have been a logistical nightmare, not to mention 401, 404, and 1600 permit compliance. If the bore and jacking operation was not implemented, it is quite possible that the permits for the project would have been revoked, and the $17 million project could have been scuttled. Manage the risk.

Know Your Permits

I was in charge of building a culvert reconstruction in an ephemeral creek in Northern California where the county entity called for the original river run material to be replaced under the bridge where the culvert was being replaced. This was a "cold, spawn, and migratory" creek that had water in it at the time of the project. I informed the engineer that this was against the Department of Fish and Wildlife Service permit specifications. The engineer ignored me, and I got him to write me a letter directing me to perform the work per plan.

It wasn't long after that that the same engineer was paying the crew extra work at force account to remove the silt- and sediment-laden original river run and replace it with imported cobble and washed river run under the watchful eye of the local game warden. This was an expensive mistake that should have never happened, precipitated by the engineer's lack of experience.

In closing, it is my hope that every person who reads this publication can say that the call to environmental stewardship awareness has been heightened. For hundreds of years, the earth has borne the result of unconscionable poisoning and pillaging, both willfully and negligently. These practices are still going on today in the United

States and around the world. It is undisputed that farming and construction activities create deleterious environmental conditions at times, but it is up to us to do what we can to control, deter, and minimize the negative environmental impacts that we *can* control. Do your part. Care.

Epilogue

As a child growing up, I spent as much time as possible outdoors in nature. Wildlife and nature are a major part of the fabric that makes me who I am. After growing up I noticed that there were no more toads in the areas where I played as a child, but there were vineyards.

In my generation I may most likely say good-bye to the polar bear and possibly the Alaskan king (Chinook) salmon and other wildlife friends. Their numbers are dwindling annually, at least for now. Whether this is caused by global warming, other natural phenomena, or poor environmental stewardship, it is happening. Let those of us who are aware be good global, corporate, and/or professional citizens and stewards of the natural resources we are entrusted with. Some things change that we have no control or responsibility over. Let us be responsible for the things that we can control and sustain.

In closing, consider the point source to receiving water risk matrix. If there is a point source of a pollutant (sediment, solid waste, hazardous waste, and deleterious material) that has a way of being conveyed (a path) to receiving water, via storm water, wind or non storm water discharge, what are the chances and means of the pollutant reaching the receiving water? Is precipitation predicted and how much. What type of soil is present and what is the runoff coefficient? What is the probability that pollutants will be discharged into the receiving water? This is the risk management matrix equation. What must we do to keep pollutants from leaving or being conveyed from point sources to receiving waters? Are you up to the challenge and do you care?

Care.

GLOSSARY OF TERMS

As found in the *California Regional Water Quality Control Board 2009 Construction General Permit: http://www.waterboards.ca.gov/water_issues/programs/stormwater/docs/constpermits/wqo_2009_0009_complete.pdf*

Active Areas of Construction

All areas subject to land surface disturbance activities related to the project including, but not limited to, project staging areas, immediate access areas, and storage areas. All previously active areas are still considered active areas until final stabilization is complete.

Active Treatment System (ATS)

A treatment system that employs chemical coagulation, chemical flocculation, or electrocoagulation to aid in the reduction of turbidity caused by fine suspended sediment.

Air Deposition

Airborne particulates from construction activities.

Approved Signatory

A person who has been authorized by the legally responsible person to sign, certify, and electronically submit permit registration documents, notices of termination, and any other documents, reports, or information required by the general permit, the state or regional water board, or USEPA.

Best Available Technology Economically Achievable (BAT)

As defined by USEPA, BAT is a technology-based standard established by the Clean Water Act (CWA) as the most appropriate means available on a national basis for controlling the direct discharge of toxic and nonconventional pollutants to navigable waters. The BAT effluent limitations guidelines, in general, represent the best existing performance of treatment technologies that are economically achievable within an industrial point source category or subcategory.

Best Conventional Pollutant Control Technology (BCT)
As defined by USEPA, BCT is a technology-based standard for the discharge from existing industrial point sources of conventional pollutants, including biochemical oxygen demand (BOD), total suspended sediment (TSS), fecal coliform, pH, oil, and grease.

Best Management Practices (BMPs)
BMPs are scheduling of activities, prohibitions of practices, maintenance procedures, and other management practices to prevent or reduce the discharge of pollutants. BMPs also include treatment requirements, operating procedures, and practices to control site runoff, spillage or leaks, sludge or waste disposal, or drainage from raw material storage.

Best Professional Judgment (BPJ)
The method used by permit writers to develop technology-based NPDES permit conditions on a case-by-case basis using all reasonably available and relevant data.

Chain of Custody (COC)
Form used to track sample handling as samples progress from sample collection to the analytical laboratory. The COC is then used to track the resulting analytical data from the laboratory to the client. COC forms can be obtained from an analytical laboratory upon request.

Coagulation
The clumping of particles in a discharge to settle out impurities, often induced by chemicals such as lime, alum, and iron salts.

Debris
Litter, rubble, discarded refuse, and remains of destroyed inorganic anthropogenic waste.

Direct Discharge
A discharge that is routed directly to waters of the United States by means of a pipe, channel, or ditch (including a municipal storm sewer system) or through surface runoff.

Discharger
The *legally responsible person* (see definition) or entity subject to a general permit.

Drainage Area
The area of land that drains water, sediment, pollutants, and dissolved materials to a common outlet.

Effluent
Any discharge of water by a discharger either to the receiving water or beyond the property boundary controlled by the discharger.

Effluent Limitation
Any numeric or narrative restriction imposed on quantities, discharge rates, and concentrations of pollutants that are discharged from point sources into waters of the United States, waters of the contiguous zone, or the ocean.

Erosion
The process by which soil particles are detached and transported by the actions of wind, water, or gravity.

Erosion Control BMPs
Vegetation, such as grasses and wildflowers, and other materials, such as straw, fiber, stabilizing emulsion, protective blankets, and so on, placed to stabilize areas of disturbed soils, reduce loss of soil due to the action of water or wind, and prevent water pollution.

Field Measurements
Testing procedures performed in the field with portable field-testing kits or meters.

Final Stabilization
All soil disturbing activities at each individual parcel within the site have been completed in a manner consistent with the requirements in a general permit.

First Order Stream

Stream with no tributaries.

Flocculants
Substances that interact with suspended particles and bind them together to form flocs.

Good Housekeeping BMPs
BMPs designed to reduce or eliminate the addition of pollutants to construction site runoff through analysis of pollutant sources, implementation of proper handling/disposal practices, employee education, and other actions.

Grading Phase (part of the Grading and Land Development Phase)
Includes reconfiguring the topography and slope, including alluvium removals, canyon cleanouts, rock undercuts, keyway excavations, land form grading, and stockpiling of select material for capping operations.

Hydromodification
Hydromodification is the alteration of the hydrologic characteristics of coastal and non-coastal waters, which in turn could cause degradation of water resources. Hydromodification can cause excessive erosion and/or sedimentation rates, causing excessive turbidity and channel aggradation and/or degradation.

Identified Organisms
Organisms within a subsample that are specifically identified and counted.

Inactive Areas of Construction
Areas of construction activity that are not active and those that have been active and are not scheduled to be disturbed again for at least fourteen days.

Index Period
The period of time during which bioassessment samples must be collected to produce results suitable for assessing the biological integrity of streams and rivers. Instream communities naturally vary over the course of a year, and sampling during the index period ensures that samples are collected during a time frame when communities are stable so that year-to-year consistency is obtained. The index period approach provides a cost-effective alternative to year-round sampling. Furthermore, sampling within the appropriate index period will yield results that are comparable to the assessment thresholds or criteria for a given region, which are established for the same index period. Because index periods differ for different parts of the state, it is essential to know the index period for your area.

K Factor

The soil erodibility factor used in the revised universal soil loss equation (RUSLE). It represents the combination of detachability of the soil, runoff potential of the soil, and the transportability of the sediment eroded from the soil.

Legally Responsible Person

The legally responsible person (LRP) will typically be the project proponent.

The categories of persons or entities that are eligible to serve as the LRP are set forth herein For any construction or land disturbance project where multiple persons or entities are eligible to serve as the LRP, those persons or entities shall select a single LRP. In exceptional circumstances, a person or entity that qualifies as the LRP may provide written authorization to another person or entity to serve as the LRP. In such a circumstance, the person or entity that provides the authorization retains all responsibility for compliance with the general permit.

Except as provided in Section II, Rational, Letter "D" Page 11 of the California General Permit, a contractor who does not satisfy the requirements is not qualified to be an LRP.

Likely Precipitation Event

Any weather pattern that is forecasted to have a 50 percent or greater chance of producing precipitation in the project area. The discharger shall obtain likely precipitation forecast information from the National Weather Service Forecast Office (e.g., by entering the zip code of the project's location at http://www.srh.noaa.gov/forecast).

Natural Channel Evolution

The physical trend in channel adjustments following a disturbance that causes the river to have more energy and degrade or aggrade more sediment. Channels have been observed to pass through five to nine evolution types. Once they pass though the suite of evolution stages, they will rest in a new state of equilibrium.

Non-Storm Water Discharges

Non-storm water discharges are discharges that do not originate from precipitation events. They can include, but are not limited to, discharges of process water, air-conditioner condensate, noncontact cooling water, vehicle wash water, sanitary wastes, concrete washout water, paint wash water, irrigation water, or pipe testing water.

Non-Visible Pollutants
Pollutants associated with a specific site or activity that can have a negative impact on water quality, but cannot be seen though observation (e.g., chlorine). Such pollutants being discharged is not authorized.

Numeric Action Level (NAL)
Level is used as a warning to evaluate if best management practices are effective and take necessary corrective actions. Not an effluent limit.

Original Sample Material
The material (i.e., macroinvertebrates, organic material, gravel, etc.) remaining after the subsample has been removed for identification.

pH
Unit universally used to express the intensity of the acid or alkaline condition of a water sample. The pH of natural waters tends to range between 6 and 9, with neutral being 7. Extremes of pH can have deleterious effects on aquatic systems.

Preliminary Phase (Preconstruction Phase—Part of the Grading and Land Development Phase)
Construction stage including rough grading and/or disking, clearing and grubbing operations, or any soil disturbance prior to mass grading.

Project-Qualified SWPPP Developer
Individual who is authorized to develop and revise SWPPPs.

Project-Qualified SWPPP Practitioner
Individual assigned responsibility for non-storm water and storm water visual observations, sampling, and analysis, and responsibility to ensure full compliance with the permit and implementation of all elements of the SWPPP, including the preparation of the annual compliance evaluation and the elimination of all unauthorized discharges.

Qualifying Rain Event
Any event that produces 0.5 inches or more of precipitation, with a forty-eight-hour or greater period between rain events.

R Factor
Erosivity factor used in the revised universal soil loss equation (RUSLE). The R factor represents the erosivity of the climate at a particular location. An average annual value of R is determined from historical weather records using erosivity values determined for individual storms. The erosivity of an individual storm is computed as the product of the storm's total energy, which is closely related to storm amount, and the storm's maximum thirty-minute intensity.

Rain Event Action Plan (REAP)
Written document, specific for each rain event, that when implemented is designed to protect all exposed portions of the site within forty-eight hours of any likely precipitation event.

Revised Universal Soil Loss Equation (RUSLE)
Empirical model that calculates average annual soil loss as a function of rainfall and runoff erosivity, soil erodibility, topography, erosion controls, and sediment controls.

Run-on
Discharges that originate off-site and flow onto the property of a separate project site.

Sampling and Analysis Plan
Document that describes how the samples will be collected under what conditions, where and when the samples will be collected, what the sample will be tested for, what test methods and detection limits will be used, and what methods/procedures will be maintained to ensure the integrity of the sample during collection, storage, shipping, and testing (i.e., quality assurance/quality control protocols).

Sediment
Solid particulate matter, both mineral and organic, that is in suspension, is being transported, or has been moved from its site of origin by air, water, gravity, or ice and has come to rest on the earth's surface either above or below sea level.

Sediment Control BMPs
Practices that trap soil particles after they have been eroded by rain, flowing water, or wind. They include those practices that intercept and slow or detain the flow of storm

water to allow sediment to settle and be trapped (e.g., silt fence, sediment basin, fiber rolls, etc.).

Sedimentation
Process of deposition of suspended matter carried by water, wastewater, or other liquids, by gravity. It is usually accomplished by reducing the velocity of the liquid below the point at which it can transport the suspended material.

Settleable Solids (SS)
Solid material that can be settled within a water column during a specified time frame. It is typically tested by placing a water sample into an Imhoff settling cone and then allowing the solids to settle by gravity for a given length of time.

Results are reported either as a volume (mL/L) or a mass (mg/L) concentration.

Sheet Flow
Flow of water that occurs overland in areas where there are no defined channels where the water spreads out over a large area at a uniform depth.

Soil Amendment
Any material that is added to the soil to change its chemical properties, engineering properties, or erosion resistance that could become mobilized by storm water.

Suspended Sediment Concentration (SSC)
The measure of the concentration of suspended solid material in a water sample by measuring the dry weight of all of the solid material from a known volume of a collected water sample. Results are reported in mg/L.

Total Suspended Solids (TSS)
The measure of the suspended solids in a water sample includes inorganic substances, such as soil particles, and organic substances, such as algae, aquatic plant/animal waste, particles related to industrial/sewage waste, and so on. The TSS test measures the concentration of suspended solids in water by measuring the dry weight of a solid material contained in a known volume of a subsample of a collected water sample. Results are reported in mg/L.

Toxicity

The adverse response(s) of organisms to chemicals or physical agents ranging from mortality to physiological responses such as impaired reproduction or growth anomalies.

Turbidity

The cloudiness of water quantified by the degree to which light traveling through a water column is scattered by the suspended organic and inorganic particles it contains. The turbidity test is reported in nephelometric turbidity units (NTU) or Jackson turbidity units (JTU).

Vertical Construction Phase

The build-out of structures from foundations to roofing, including rough landscaping.

Water Quality Objectives (WQO)

Water quality objectives are defined in the California Water Code as limits or levels of water quality constituents or characteristics, which are established for the reasonable protection of beneficial uses of water or the prevention of nuisance within a specific area.

Waters of the United States

Generally refers to surface waters, as defined by the federal Environmental Protection Agency in 40 C.F.R. § 122.2.

ACRONYMS

As found in the *California Regional Water Quality Control Board 2009 Construction General Permit at http://www.waterboards.ca.gov/water_issues/programs/stormwater/ docs/constpermits/wqo_2009_0009_complete.pdf*

These acronyms are California-centric, and some of them may not apply to the state or area in which you live; they are for information only.

ASBS	Areas of Special Biological Significance
ASTM	American Society of Testing and Materials—Standard test method for particle-size analysis of soils
ATS	Active Treatment System
BASMAA	Bay Area Storm water Management Agencies Association
BAT	Best Available Technology Economically Achievable
BCT	Best Conventional Pollutant Control Technology
BMP	Best Management Practices
BOD	Biochemical Oxygen Demand
BPJ	Best Professional Judgment
CAFO	Confined Animal Feeding Operation
CCR	California Code of Regulations
CEQA	California Environmental Quality Act
CFR	Code of Federal Regulations
CGP NPDES	General Permit for Storm Water Discharges Associated with Construction Activities
CIWQS	California Integrated Water Quality System
CKD	Cement Kiln Dust
COC	Chain of Custody
CPESC	Certified Professional in Erosion and Sediment Control
CPSWQ	Certified Professional in Storm Water Quality
CSMP	Construction Site Monitoring Program
CTB	Cement Treated Base
CTR	California Toxics Rule
CWA	Clean Water Act
CWC	California Water Code

CWP	Center for Watershed Protection
DADMAC	Diallyldimethyl-ammonium chloride
DDNR	Delaware Department of Natural Resources
DFG	Department of Fish and Game
DHS	Department of Health Services
DWQ	Division of Water Quality
EC	Electrical Conductivity
ELAP	Environmental Laboratory Accreditation Program
EPA	Environmental Protection Agency
ESA	Environmentally Sensitive Area
ESC	Erosion and Sediment Control
HSPF	Hydrologic Simulation Program Fortran
JTU	Jackson Turbidity Units
LID	Low Impact Development
LOEC	Lowest Observed Effect Concentration
LRP	Legally Responsible Person
LUP	Linear Underground/Overhead Projects
MATC	Maximum Allowable Threshold Concentration
MDL	Method Detection Limits
MRR	Monitoring and Reporting Requirements
MS4	Municipal Separate Storm Sewer System
MUSLE	Modified Universal Soil Loss Equation
NAL	Numeric Action Level
NEL	Numeric Effluent Limitation
NICET	National Institute for Certification in Engineering Technologies
NOAA	National Oceanic and Atmospheric Administration
NOEC	No Observed Effect Concentration
NOI	Notice of Intent
NOT	Notice of Termination
NPDES	National Pollutant Discharge Elimination System
NRCS	Natural Resources Conservation Service
NTR	National Toxics Rule
NTU	Nephelometric Turbidity Units
O&M	Operation and Maintenance
PAC	Polyaluminum Chloride
PAM	Polyacrylamide

PASS	Polyaluminum Chloride Silica/Sulfate
PCC	Portland Cement Concrete
POC	Pollutants of Concern
PoP	Probability of Precipitation
POTW	Publicly Owned Treatment Works
PRDs	Permit Registration Documents
PWS	Planning Watershed
QAMP	Quality Assurance Management Plan
QA/QC	Quality Assurance/Quality Control
REAP	Rain Event Action Plan
RWQCB	Regional Board Regional Water Quality Control Board
ROWD	Report of Waste Discharge
RUSLE	Revised Universal Soil Loss Equation
RW	Receiving Water
SMARTS	Storm water Multi Application Reporting and Tracking System
SS	Settleable Solids
SSC	Suspended Sediment Concentration
SUSMP	Standard Urban Storm Water Mitigation Plan
SW	Storm Water
SWAMP	Surface Water Ambient Monitoring Program
SWARM	Storm Water Annual Report Module
SWMM	Storm Water Management Model
SWMP	Storm Water Management Program
SWPPP	Storm Water Pollution Prevention Plan
TC	Treatment Control
TDS	Total Dissolved Solids
TMDL	Total Maximum Daily Load
TSS	Total Suspended Solids
USACOE	US Army Corps of Engineers
USC	United States Code
USEPA	United States Environmental Protection Agency
USGS	United States Geological Survey
WDID	Waste Discharge Identification Number
WDR	Waste Discharge Requirements
WET	Whole Effluent Toxicity
WLA	Waste Load Allocation

WQBEL	Water Quality Based Effluent Limitation
WQO	Water Quality Objective
WQS	Water Quality Standard
WRCC	Western Regional Climate Center

REFERENCES AND SOURCES

1. *California Regional Water Quality Control Board Construction General Permit 2009-0009-DWQ amended by 2010-0014-DWQ & 2012-0006-DWQ.*

2. *California Department of Transportation 2010 Standard Specifications.*

3. *California Department of Transportation 2010 Standard Plans.*

4. *EnviroCert International Certified Professional in Erosion and Sediment Control exam review study guide, 2010.*

5. Wikipedia Clean Water Act Point Sources definition: http://en.wikipedia.org/wiki/Clean_Water_Act.

6. Silt fence installation detail: http://www.elibrary.dep.state.pa.us.

7. Check dam spacing detail: http:www.imgkid.com.

8. Straw bale barrier detail: http://www.google.com/urFextension.missouri.edu.

9. Drain inlet protection: http://www.google.com/www.biocleanenvironmental.com.

10. Temporary tire wash detail: www.rainforent.com.

11. City and County of Honolulu, November 2011 Storm Water Best Management Practice Manual by the City and County of Honolulu Department of Environmental Services.

The Author, Mike Peters.

"The environment and its original, organic constituents that we see today are but a reminder of what was once an amazing creation wrought by the synergistic union of the collaboration of time and space as one...

Let us sustain what we can as responsible, caring stewards of an invaluable and irreplaceable resource."

—Mike Peters,

Author of *Storm Water Pollution Prevention Plan Risk Management* and founder of the Envirprotechnicalgroup: Environmental Professionals for Effective Sustainable Environmental Protection and Best Management Practice Implementation, http://www.envirprotechnicalgroup.com.

About the Author

Mike L. Peters
QSD, CPESC

M ike L. Peters is a Certified Professional in Erosion and Sediment Control (CPESC) and a Qualified Storm Water Pollution Prevention Plan Developer (QSD) in the State of California as registered with the California Storm Water Quality Association (CASQA) and EnviroCert International. He is founder of www.envirprotechnicalgroup.com, a website for environmental professional services.

For the last six years, Mike has performed Independent Assurance (IA) audits of Storm Water Pollution Prevention Plans (SWPPPs) and Water Pollution Control Plans (WPCPs) as an independent-assurance third-party consultant.

His experiences from conducting hundreds of independent-assurance audits and inspections of construction-site SWPPP and WPCP projects and maintenance facilities are in this publication. Understanding the risk management factors of project management, project ownership, and permit compliance can make the difference between the success and failure of a project, both economically and financially. Environmental stewardship and responsibility is upon us all at this time. Let us implement the best management practices, means, and methods available to preserve the natural environmental assets and treasures left to our responsibility.

Mike enjoys fishing in Alaska every year and making fine wines in the Napa/ Sonoma County regions of Northern California.